Measurement Issues in Aging and Physical Activity

Proceedings of the 10ᵗʰ Measurement and Evaluation Symposium

Weimo Zhu, PhD

University of Illinois at Urbana-Champaign

Wojtek Chodzko-Zajko, PhD

University of Illinois at Urbana-Champaign

Editors

Human Kinetics

Library of Congress Cataloging-in-Publication Data

Measurement and Evaluation Symposium (10th : 2003 : University of Illinois at Urbana-Champaign)
    Measurement issues in aging and physical activity / [edited by] Weimo Zhu, Wojtek Chodzko-Zajko.
        p.   ;   cm.
    Selected papers from the 10th Measurement and Evaluation Symposium, held Oct. 16-18, 2003 at University of Illinois at Urbana-Champaign.
    Includes bibliographical references.
    ISBN 0-7360-5364-6 (hard cover)
    1. Aging--Physiological aspects--Congresses.   2. Exercise--Physiological aspects--Congresses.   3. Exercise for older people--Physiological aspects--Congresses.
    [DNLM: 1. Aging-physiology--Congresses.   2. Exertion--physiology--Aged--Congresses. WT 104 M4855m 2006]   I. Zhu, Weimo, 1955-   II. Chodzko-Zajko, Wojtek J.   III. Title.
QP86.M43 2006
612.6'7--dc22

                                           2005017703
ISBN: 0-7360-5364-6

The Web addresses cited in this text were current as of *May 17, 2005,* unless otherwise noted.

Acquisitions Editor: Loarn D. Robertson, PhD; **Managing Editor:** Lee Alexander; **Copyeditor:** Joyce Sexton; **Proofreader:** Erin Cler; **Permission Manager:** Dalene Reeder; **Graphic Designer:** Robert Reuther; **Graphic Artist:** Kathleen Boudreau-Fuoss; **Photo Manager:** Sarah Ritz; **Cover Designer:** Keith Blomberg; **Photographer (interior):** Photos on p. 53 courtesy of Uriel Cohen. Photos on p. 108 courtesy of Miyoung Lee; **Art Manager:** Kelly Hendren; **Illustrators:** Kelly Hendren and Al Wilborn; **Printer:** Edwards Brothers

Printed in the United States of America   10 9 8 7 6 5 4 3 2 1

Human Kinetics
Web site: www.HumanKinetics.com

United States: Human Kinetics, P.O. Box 5076, Champaign, IL 61825-5076
800-747-4457
e-mail: humank@hkusa.com

Canada: Human Kinetics
475 Devonshire Road Unit 100, Windsor, ON N8Y 2L5
800-465-7301 (in Canada only)
e-mail: orders@hkcanada.com

Europe: Human Kinetics
107 Bradford Road, Stanningley, Leeds LS28 6AT, United Kingdom
+44 (0) 113 255 5665
e-mail: hk@hkeurope.com

Australia: Human Kinetics
57A Price Avenue, Lower Mitcham, South Australia 5062
08 8277 1555
e-mail: liaw@hkaustralia.com

New Zealand: Human Kinetics
Division of Sports Distributors NZ Ltd., P.O. Box 300 226 Albany, North Shore City, Auckland
0064 9 448 1207
e-mail: info@humankinetics.co.nz

Measurement Issues in Aging and Physical Activity

Proceedings of the 10th Measurement and Evaluation Symposium

To our parents

Bing-Qi Li and Jing-Rui Zhu
Ewa and Witold Chodzko-Zajko

For their love and support

Contents

Part II   Measurement Challenges in Aging Research

Part IV   Measurement in Kinesiology: Past, Present, and Future

Sponsors

American Association for Active Lifestyles and Fitness

American College of Sports Medicine

Centers for Disease Control and Prevention

University of Illinois at Champaign-Urbana

Human Kinetics, Inc.

Active for Life

National Blueprint: Increasing Physical Activity among Adults Aged 50 or Older

Mettler Center for Whole Life Fitness

Carle Therapy Services

Dynastream Innovations

CIR Systems, Inc.

University of Illinois Campus Recreation

Ambulatory Monitoring, Inc.

American Physical Therapy Association

Body Media, Inc.

Cambridge Isotope Labs, Inc.

Mini-Mitter Co., Inc.

Partnership for Prevention

Contributors

Ted A. Baumgartner
University of Georgia, Athens, GA

Cynthia Castro, PhD
Stanford University School of Medicine,
Stanford, CA

Wojtek Chodzko-Zajko, PhD
University of Illinois at Urbana-Champaign,
Urbana, IL

Ruth Cohen, PhD
Cohen Consultants, Milwaukee, WI

Uriel Cohen, DArch
University of Wisconsin, Milwaukee, WI

Diane Dowdy, PhD
The Texas A&M University System
Health Science Center,
School of Rural Public Health,
College Station, TX

Steriani Elavsky
University of Illinois at Urbana-Champaign,
Urbana, IL

Russell Glasgow, PhD
Kaiser Permanente, Penrose, CO

Michele Guerra, MS, CHES
Human Kinetics, Champaign, IL

Leigh W. Jerome, PhD
Pacific Telehealth & Technology Hui,
Honolulu, HI

Caroline J. Ketcham
Texas A&M University, College Station, TX

Abby King, PhD
Stanford University School of Medicine,
Stanford, CA

Laura Leviton, PhD
The Robert Wood Johnson Foundation,
Princeton, NJ

Edward McAuley
University of Illinois at Urbana-Champaign,
Urbana, IL

Roderick P. McDonald
University of Illinois at Urbana-Champaign,
Urbana, IL

Robin Mockenhaupt, PhD, MPH
The Robert Wood Johnson Foundation,
Princeton, NJ

Dale P. Mood, PhD
University of Colorado, Boulder, CO

James R. Morrow, Jr., PhD
University of North Texas, Denton, TX

Miriam E. Nelson, PhD, FACSM
Director, The John Hancock Center
for Physical Activity and Nutrition
The Gerald J. and Dorothy R. Friedman
School of Nutrition Science and Policy,
Tufts University, Boston, MA

Marcia Ory, PhD, MPH
The Texas A&M University System
Health Science Center,
School of Rural Public Health,
College Station, TX

Tuomo Rankinen, PhD
Human Genomics Laboratory,
Pennington Biomedical Research Center,
Baton Rouge, LA

James H. Rimmer, PhD
Department of Disability and Human Development
University of Illinois at Chicago, Chicago, IL

Brigid Sanner
Sanner & Company, Dallas, TX

Rebecca Seguin, MS, CSCS
*The John Hancock Center for Physical Activity
and Nutrition
The Gerald J. and Dorothy R. Friedman
School of Nutrition Science and Policy,
Tufts University, Boston, MA*

Roy J. Shephard
*Faculty of Physical Education and Health,
Department of Public Health Sciences,
Faculty of Medicine, University of Toronto,
Toronto, ON*

Stephen Silverman, EdD
*Teachers College,
Columbia University, New York, NY*

George E. Stelmach
*Motor Control Laboratory, Arizona State University,
Tempe, AZ*

Sara Wilcox, PhD
University of South Carolina, Columbia, SC

Weimo Zhu, PhD
*University of Illinois at Urbana-Champaign,
Urbana, IL*

Preface

The 10th M&E Symposium, "Measurement Issues and Challenges in Aging Research," held at the University of Illinois at Urbana-Champaign on October 16 through 18, 2003, was an extraordinary success. More than 140 aging and measurement researchers and graduate students from around the world participated in the symposium, and the feedback received was extremely positive.

There were several unique features to the 10th M&E Symposium. First, a distinguished panel of world-class scholars in the fields of aging and measurement presented the latest issues, challenges, and techniques for their respective fields. Second, a preconference workshop was included in the program for the first time. In the workshop, two advanced statistical and research methodologies were introduced: structural equation modeling and longitudinal data analysis. Both sessions were well attended. Third, very broad topics of aging research were covered, including kinesiology, biology, physiology, psychology, technology, culture, urban planning, and statistics. Many state-of-the-art research fields (e.g., human gene mapping, remote clinical assessment, and online computerized assessment) were discussed, and future trends and research directions were addressed. Fourth, multicultural factors associated with physical activity promotion and physical activity interventions were discussed by several speakers. In addition, measurement issues and challenges in alternative medicine were addressed. Master Ping-Zhang Chen from China gave an astounding demonstration of Qi-gong, a traditional Chinese exercise, in which he showed how Qi, or energy force, can be used to move a person apparently without physical contact. Finally, issues related to measurement training for aging researchers and training for future measurement specialists were discussed at the symposium. The adoption of a new term, "kinesmetrics," as a label for the field of measurement and evaluation, was discussed and was generally supported. Kinesmetrics is defined as "a discipline to develop and apply measurement theory, statistics, and mathematical analysis to the field of kinesiology."

As hoped, the symposium provided a unique forum for both aging researchers and measurement specialists to discuss measurement issues in aging and physical activity research. Many new researchers and graduate students were introduced to the field and motivated to begin to address some of the research issues explored at the symposium. This is perhaps the most important outcome of the symposium. The impact of the symposium on the future of aging research and kinesmetrics will be substantial.

This volume includes 14 papers presented at the symposium that are organized into four sections. There are five chapters in part I, "Critical Issues in Aging and Physical Activity Research."

• In chapter 1, Rankinen provides a timely update on progress in the area of human gene mapping, physical activity/fitness, and aging. He notes that the completion of the Human Genome Project has opened new possibilities to understanding physical fitness, activity, and aging at the molecular and genetic levels, especially with respect to the biological mechanisms contributing to interindividual differences in responding to physical activity and exercise training.

• In chapter 2, Nelson and Seguin provide a comprehensive review of exercise and older adults, evaluating exercise protocols in a variety of populations. Their conclusion is exciting: A variety of exercise prescriptions, from highly controlled laboratory trials to minimally supervised home-based programs, have demonstrated the ability to elicit meaningful health benefits in older adults. The authors call for more research to identify the most appropriate recommendations for older adults and for increased access to safe and effective programs in various settings.

• In chapter 3, Shephard addresses the exercise dose-response effect, one of the most important issues in exercise prescription for older adults. After providing an overview on research in this area, Shephard presents several critical issues in the dose-response research for older adults. For example, Is aerobic fitness enough? Should we prescribe moderate- or high-intensity physical activity? In addition, specific recommendations for individuals with health problems (e.g., stroke, obesity, diabetes) are reviewed in detail. Finally, future research priorities are outlined.

• In chapter 4, Ketcham and Stelmach provide a comprehensive review on the control and regulation of movement in older adults. Understanding how movement in older adults is produced and what mechanisms are involved in the control and regulation of the movement is essential. As new recording and analysis methods are introduced, much better insight has been achieved with respect to how the control and regulation of movement decline with aging. After a general overview of major research issues and studies, the authors provide a detailed discussion of movement characteristics (i.e., movement duration, kinematic profiles, force control and regulation, and coordination), skill learning, and visual monitoring. Future research directions are also addressed.

• In chapter 5, the final chapter in part I, Cohen and Cohen look at two critical yet often overlooked areas related to older adults' physical activities: environment and culture. As they point out, to promote physical activities among older adults, the barriers encountered by older adults residing in active living environments must be identified and then removed, and cultural heritage should be used as a catalyst for active living. Illustrating with a case study, Cohen and Cohen propose using culture-based physical activity as an alternative approach for active living promotion.

Part II, "Measurement Challenges in Aging Research," consists of five chapters.

- In chapter 6, McAuley and Elavsky examine issues in physical activity, aging, and quality of life (QOL), with a focus on measurement. After providing a general review on QOL, they address critical topics in this emerging research area, such as measuring QOL in physical activity research, the impact of physical activity levels of older adults on QOL, and the dose-response effect and QOL. Finally, related measurement issues are examined in detail.

- In chapter 7, Rimmer examines assessment issues related to physical activity and disability. When persons with disabilities age, they often experience more physical challenges. After a brief description of disability research, Rimmer addresses measurement issues in more depth, including heterogeneity of disability, specific health outcomes, surveillance, self-report measures, and functional assessment.

- In chapter 8, Morrow and Mood describe a new and exciting research area: the impact of the environment on physical activity in older adults. After briefly defining four dimensions of the environment that may have an impact on older adults' physical activities, they address key measurement and research issues in environmental research.

- In chapter 9, Ory and her colleagues address measurement and aging research from a different angle. As they point out, most research-based interventions are not tested in real-world settings, instead focusing on maximizing internal validity. These interventions often have poor external validity and are difficult to translate into real-life settings. By underscoring the need to examine real-world situations in the development of future research models, the authors describe both the status of current research and future research needs.

- In the last chapter of this part, chapter 10, Zhu provides an overview of Qi, aging, and measurement. Master Chen's demonstration and Dr. Shan Wong's talk on Qi and Qi-gong at the conference brought many questions, including skeptical ones: Is there really some human ability we don't know about or is it simply trickery? Along with introducing the history, mystery, and controversy associated with Qi-gong, the chapter underscores the need for additional measurement research in this area.

Part III consists of two chapters, with a focus on "New Measurement Methods and Techniques." Repeated measures and longitudinal study design are commonly used in aging research. Traditionally used statistical methods, for example repeated measures of ANOVA, are known for their limitations (e.g., requirement of the same time interval and difficulty in handling missing data). Several trend/growth curve models have been brought to our attention recently.

- In chapter 11, McDonald provides an easy-to-understand tutorial on trend/growth curve models, and it is expected that these models will become major analytic tools in aging research practice in the coming years.

• Another dramatic change in our lives is the rapid development of new technology. The era of *e*-health is coming! In chapter 12, Jerome provides an overview on emergent technologies, especially remote clinical assessment. It is clear that new developments in bioengineering are changing assessment practice in biomedical research. It is expected that there will be many applications of these assessment procedures for aging and physical activity research in the very near future.

Part IV, "Measurement in Kinesiology: Past, Present, and Future," examines the future of measurement and evaluation as a subdiscipline of kinesiology. It has been almost 30 years since the first Measurement and Evaluation Symposium was held at Indiana University in 1975. The measurement and evaluation field as a subdiscipline in kinesiology is at a critical juncture: Many pioneers in the measurement and evaluation field have retired; jobs for measurement specialists are decreasing; and only two doctoral programs are now active in North America.

• In chapter 13, Baumgartner provides a review of the work of the Measurement and Evaluation Council. After an overview of the council's past and present activity, he calls for the council to expand its scope and influence to cover the wide variety of areas in kinesiology in which specialized knowledge about measurement and evaluation is greatly needed.

• Finally, in chapter 14, Silverman provides his perspective on the future of the field of measurement and evaluation. After offering a historical review of the changing face of the measurement specialist, he challenges the field with regard to future measurement and evaluation needs by posing some critical questions: How and by whom will measurement and evaluation classes be taught? How do we help students learn about research methods? Where will measurement specialists be employed? How will kinesiology faculty be trained in measurement and evaluation? To keep measurement and evaluation active and alive, the field of kinesiology as a whole has to take action now to address these questions before it is too late.

The success of the symposium and publication of the Proceedings could not have happened without the support of many organizations and individuals. We want to take this opportunity to thank our sponsoring organizations—American Association for Active Lifestyle and Fitness (AAALF), American College of Sports Medicine (ACSM), Measurement and Evaluation Council, University of Illinois at Urbana-Champaign, Human Kinetics (HK), the symposium sponsors, and the members of the organization and scientific committees—for their full support. We especially appreciate Jane Senior and Amy Katzenberger at ACSM and Janet Seaman at AAALF for their hard work and timely assistance. Finally, we thank Loarn Robertson and Lee Alexander at HK for their skillful and patient assistance on the publication of the Proceedings.

Part I

Critical Issues in Aging and Physical Activity Research

Chapter 1

Human Gene Map, Physical Activity, and Aging

Tuomo Rankinen, PhD

The last decade has witnessed amazing advances in molecular genetics. The completion of the Human Genome Project has opened completely new possibilities for exploring the molecular and genetic basis of complex traits, and it holds great promise for the development of new insights into biological mechanisms contributing to interindividual differences in responsiveness to acute exercise and exercise training. The purpose of this review is to briefly summarize the current status of research concerned with the genetics of health-related fitness and discuss some specific features associated with aging.

Genetics and Health-Related Fitness

The favorable effects of endurance training on physical performance and on several risk factors for chronic diseases have been well documented. However, it is also clear that there are marked differences between individuals in the adaptation to exercise training. For example, in the HERITAGE Family Study, 742 healthy but sedentary subjects followed an identical, well-controlled endurance training program for 20 weeks. Although they were subjected to the same training stimulus, increases in maximal oxygen consumption ($\dot{V}O_2$max), a measure of cardiorespiratory fitness, varied from no change to increases of more than 1 L/min (Bouchard & Rankinen, 2001; Skinner et al., 2000). A similar pattern of variation in training responses was observed for several other phenotypes, such as plasma lipid levels and submaximal exercise heart rate and blood pressure (Leon et al., 2000; Bouchard & Rankinen, 2001; Wilmore et al., 2001). These data emphasize that the effects of endurance training on cardiovascular traits should be evaluated not only in terms of mean changes but also in terms of response heterogeneity. Furthermore, high and low responsiveness to training in the HERITAGE Family Study was characterized by significant familial aggregation; that is, some families were mainly low responders whereas in some families all the members showed significant improvements. These observations support the notion that individual variability is a normal biological phenomenon, which may reflect genetic diversity (Bouchard, 1995; Bouchard & Rankinen, 2001).

Genetic Epidemiology

The maximal heritability of a trait, that is, the combined effect of genes and shared environment on a phenotype, can be estimated using data from family and twin studies. The heritability estimates are based on comparisons of phenotypic similarities between pairs of relatives with different level of biological relatedness. For example, biological siblings, who share about 50% of their genes identical by descent (IBD), should be phenotypically more similar than their parents (biologically unrelated individuals) if genetic factors contribute to the trait of interest. Likewise, a greater phenotypic resemblance between identical twins (100% of genes IBD) than between dizygotic twins (50% of genes IBD) indicates genetic effect on the phenotype.

The heritability of $\dot{V}O_2max$ in the sedentary state has been estimated from a few twin and family studies, the most comprehensive of these being the HERITAGE Family Study (Bouchard et al., 1998). An analysis of variance revealed a clear familial aggregation of $\dot{V}O_2max$ in the sedentary state, and family lines accounted for about 40% of the variance in $\dot{V}O_2max$. Maximum likelihood estimation of familial correlations revealed a maximal heritability of 51% for $\dot{V}O_2max$ (Bouchard et al., 1998). Data from the twin studies have yielded very similar heritability estimates, ranging from 25% to 66% (Bouchard et al., 1986; Fagard, Bielen, & Amery, 1991; Sundet, Magnus, & Tambs, 1994).

Among pairs of monozygotic twins, the $\dot{V}O_2max$ response to standardized training programs showed six to nine times more variance between genotypes (between pairs of twins) than within genotypes (within pairs of twins) (Bouchard et al., 1992). Thus, gains in absolute $\dot{V}O_2max$ were much more heterogeneous between pairs of twins than within pairs of twins. In the HERITAGE Family Study, the increase in $\dot{V}O_2max$ in 481 individuals from 99 two-generation families of Caucasian descent showed 2.6 times more variance between families than within families, and the model-fitting analytical procedure yielded a maximal heritability estimate of 47% (Bouchard et al., 1999).

In addition to $\dot{V}O_2max$, the HERITAGE Family Study investigators examined heritability of training-induced changes in several other phenotypes, such as submaximal aerobic performance (Perusse et al., 2001); resting and submaximal exercise blood pressure, heart rate, stroke volume, and cardiac output (An et al., 2000a; An et al., 2000b; Rice et al., 2002a; An et al., 2003); body composition and body fat distribution (Rice et al., 1999; Perusse et al., 2000); and plasma lipid, lipoprotein, and apolipoprotein levels (Rice et al., 2002b). The maximal heritabilities for these traits ranged from 25% to 55%, further confirming the contribution of familial factors to the person-to-person variation in responsiveness to endurance training.

Molecular Genetics

The evidence from the genetic epidemiology studies suggests that there is a genetically determined component affecting exercise-related phenotypes. However, since these traits are complex and multifactorial in nature, the search for genes and mutations responsible for the genetic regulation must not only target several families of phenotypes, but also consider the phenotypes in the sedentary state and in response to exercise training. It is also obvious that the research on molecular genetics of exercise-related phenotypes is still in its infancy. The 2002 update of the human gene map for performance and health-related fitness included 90 gene entries and quantitative trait loci (QTL) on the autosomes and two on the X chromosome (Perusse et al., 2003) for physical performance (cardiorespiratory endurance, elite endurance athlete status, muscle strength, other muscle performance traits, exercise intolerance) and health-related fitness (hemodynamic traits, anthropometry, and body composition; insulin and glucose metabolism; blood lipids and lipoproteins and hemostatic factors) phenotypes. As a comparison, the latest version of a similar map for obesity-related phenotypes included more than 300 loci (Chagnon et al., 2003). These numbers demonstrate that relatively little has been accomplished to date. For instance, no gene contributing to human variation in endurance performance has even been identified yet as a result of studies based on model organisms. Now that we have entered the era in which large fractions of the human, mouse, and rat genomic sequences have become available, the field of exercise science and sports medicine will need to devote more attention to molecular and genetic research.

Genetic Research Approaches: Candidate Genes Versus Genomic Linkage Scans

The majority of the exercise-related molecular genetic studies published so far have utilized a *candidate gene* approach; that is, a gene has been targeted based on its potential physiological and metabolic relevance to the trait of interest. Statistical tests for an association are based on the comparison of allele and genotype frequencies of genetic markers between two groups of subjects, one having the phenotype of interest (e.g., high $\dot{V}O_2$max or endurance athletes, i.e., the "cases"), the other one not (the "controls"). However, with continuous traits, the test is done through comparison of mean phenotypic values across genotype groups, or between carriers and noncarriers of a specific allele.

An alternative strategy to identify genes affecting performance-related phenotypes relies on *linkage analysis.* The basic idea of genetic linkage is to test if a genetic locus is transmitted from one generation to the next together with a trait (or another genetic locus) of interest. The process is fairly straightforward when the trait is influenced by only one (major) gene. In these cases, the underlying genetic architecture can be deduced through observation of the transmission of the trait in affected families, which allows generating powerful models for linkage testing with genetic markers. However, multifactorial and oligogenic traits such as performance phenotypes rarely follow a specific inheritance model. In this case, it is not possible to use traditional parametric or model-based linkage methods. Instead, the linkage test for oligogenic traits is based on the idea that a pair of relatives (usually siblings) who are genetically similar should also be alike in terms of phenotypic values. The genetic similarity of the pair is determined by estimation of how many common alleles at a given locus the individuals have inherited from the same ancestors (allele sharing IBD). For more details regarding linkage methods, see Rao and Province, 2001.

The major conceptual difference between linkage and association is that association targets a specific allele or a genotype at a given gene locus whereas linkage refers to a chromosomal region rather than a specific gene or mutation. Thus, the linkage analysis can be used to identify chromosomal regions that harbor gene(s) affecting the phenotype, even if there is no a priori knowledge of the existence of such genes. By definition, the linkage analysis always requires family data. Therefore, data collection is more challenging than in case-control and cohort studies with unrelated subjects. The usefulness of linkage analysis to study elite athletes is predictably quite limited. However, in studies on inter-individual differences in responsiveness to acute exercise and exercise training, the family design can provide exploratory opportunities that are not available in association studies with independent subjects.

The identification of a QTL is only the first step in the gene discovery process. Since linkage analysis provides information about a genomic region, a typical QTL may span over several millions of base pairs. Such a vast region may contain dozens or even hundreds of genes. The procedure commonly used to identify the causal gene(s) within a QTL is called positional cloning. Briefly, the initial detection of the QTL is followed by the addition of more microsatellite markers within the QTL region with a view to narrowing down the region as much as possible using linkage analysis. Once the resolution limit of the linkage approach is reached, the next step involves fine mapping using association analyses with single nucleotide polymorphisms (SNPs). The rationale for fine mapping is that the greater number of SNPs and the greater sensitivity of the association tests enable one to derive more detailed information about the region. Even if the SNPs used for fine mapping do not include the specific DNA sequence variant(s) affecting the trait, they likely provide useful leads regarding the underlying causal mutation(s). The closely located yet functionally neutral SNPs may cosegregate with the trait-influencing DNA variant and thereby be part of the haplotype that also contains the causal allele. This phenomenon is called allelic association or linkage disequilibrium, and it forms the theoretical basis for association mapping.

Once the results from the association analyses are deemed strong enough, the next step is to screen by sequencing the candidate genes for DNA sequence variation. The confirmation of the relevance of the detected mutations includes additional association studies in the original and other populations, as well as functional assays in vitro (expression studies in different cell lines) and in vivo (transgenic and knockout animal models). The positional cloning approach has been used successfully to identify genes causing several diseases, such as long QT syndrome 1 (Wang et al., 1996) and autosomal dominant familial polymorphic ventricular tachycardia (Swan et al., 1999; Laitinen et al., 2001). Both traits are characterized by increased incidence of cardiac events during exercise.

Current Status of the Human Gene Map for Performance and Health-Related Fitness

The genetic data hold great promise for helping us understand why some individuals respond favorably to exercise training in terms of reduction of chronic disease risk factor levels whereas others do not. The 2002 update of the human gene map for performance and health-related fitness included 16 genes, from 22 studies, that had been investigated in relation to exercise training-induced changes in hemodynamic (8 genes, 11 studies), body composition (8 genes, 8 studies), plasma lipid and lipoprotein (2 genes, 2 studies), and hemostatic (3 genes, 3 studies) phenotypes (see table 1.1). The genes associated with body composition, plasma lipid, and hemostatic phenotype training responses were all based on a single study. However, with hemodynamic phenotypes, some candidate gene findings have been replicated in at least two studies. For example, an association between blood pressure training response and the angiotensinogen (AGT) M235T polymorphism has been reported in both the HERITAGE Family Study and the DNASCO study (Rankinen et al., 2000; Rauramaa et al., 2002). In white HERITAGE males, the AGT M235M homozygotes showed the greatest reduction in submaximal exercise diastolic blood pressure following a 20-week endurance training program (Rankinen et al., 2000). Likewise, in middle-aged Eastern Finnish men, the M235M homozygotes had the most favorable changes in resting systolic and diastolic blood pressure during a six-year exercise intervention trial (Rauramaa et al., 2002).

Genetics, Aging, and Health-Related Fitness

Average life expectancy in the industrialized countries has increased dramatically during the last century. It has been predicted that the proportion of the population aged 60 and above will double over the next 100 years, and it has been estimated that these age groups will compose 45% to 50% of the population by the year 2100 (Lutz, Sanderson, & Scherbov, 2001). The increase in

TABLE 1.1 The 2002 Update of the Human Gene Map for Performance and Health-Related Fitness

Gene name	Abbreviation	Location
HEMODYNAMICS		
Angiotensinogen	AGT	1q42-q43
Nitric oxide synthase 3 (endothelial cell)	NOS3	7q36
Lipoprotein lipase	LPL	8p22
Guanine nucleotide binding protein (G protein), beta polypeptide 3	GNB3	12p13
Bradykinin receptor B2	BDKRB2	14q32.1-q32.2
Angiotensin I converting enzyme	ACE	17q23
Apolipoprotein E	APOE	19q13.2
Peroxisome proliferative activated receptor, alpha	PPARA	22q13.31
BODY COMPOSITION		
Peroxisome proliferative activated receptor, gamma	PPARG	3p25
Lipoprotein lipase	LPL	8p22
Beta-3-adrenergic receptor	ADRB3	8p12-p11.2
Uncoupling protein 3	UCP3	11q13
Guanine nucleotide binding protein (G protein), beta polypeptide 3	GNB3	12p13
Vitamin D (1,25-dihydroxyvitamin D3) receptor	VDR	12q12-q14
Insulin-like growth factor 1	IGF1	12q22-q23
Angiotensin I converting enzyme	ACE	17q23
LIPIDS AND LIPOPROTEINS		
Lipoprotein lipase	LPL	8p22
Apolipoprotein E	APOE	19q13.2
HEMOSTATIC FACTORS		
Fibrinogen, A alpha polypeptide	FGA	4q28
Fibrinogen, B beta polypeptide	FGB	4q28
Plasminogen activator inhibitor 1	PAI1	7q21.3-q22

Summary of the candidate genes that have shown statistically significant associations with health-related fitness training response phenotypes in humans. For further details see Perusse et al. (2003).

life expectancy is clearly related to the improvements in health care during the 20th century, and survival to old age is known to have a strong environmental and behavioral component. Although longevity is a complex multifactorial trait, several lines of evidence also support the contribution of genetic factors to the variation of human life span, especially to the likelihood of reaching exceptionally old age. For example, death rates at all ages among siblings of centenarians are about one-half the national level in the United States; and the female and male siblings are at least 8 and 17 times, respectively, more likely to reach age 100 themselves as compared to the individuals from the U.S. 1900 birth cohort (Perls et al., 2002). Furthermore, a QTL for exceptional longevity was recently reported on chromosome 4q24-q25 in humans (Puca et al., 2001).

The growth of the elderly population presents several challenges to exercise physiology and genetics of health-related fitness. With the aging of the population, the prevalence of age-related chronic diseases is expected to increase as well. However, physical activity has been shown to be beneficial in the primary and secondary prevention of these health problems in middle-aged subjects, and the same pattern seems to apply to older individuals as well (American College of Sports Medicine, 1998). From the genetics viewpoint, aging is associated with changes in some genotype frequency distributions and with changes in gene expression patterns. The first phenomenon seems to be particularly relevant to those genotypes that predispose to conditions associated with early mortality (demographic selection). For example, the epsilon 4 allele of the apolipoprotein E gene, which has been shown to be associated with increased risk of cardiovascular disease and Alzheimer's disease, has been shown to be less frequent among centenarians than in 20- to 70-year-old controls (Schachter et al., 1994). Also the level of gene expression changes with aging. For example, messenger RNA (ribonucleic acid) of genes encoding several mitochondrial proteins involved in electron transport and ATP (adenosine triphosphate) synthesis was less abundant in skeletal muscle samples from 66- to 77-year-old individuals as compared to samples from 21- to 24-year-old subjects (Welle, Bhatt, & Thornton, 2000).

The relevance of the aging-induced changes in gene expression profile to health-related fitness traits is still poorly understood, but data from animal models indicate that regular physical activity may prevent at least some of these changes. For example, Bronikowski and coworkers (Bronikowski et al., 2003) investigated cardiac expression levels of 11,904 genes in middle-aged and old male mice derived from sedentary and spontaneously physically active selective breeding lines. In the sedentary animals, 137 genes showed significant aging-related changes in expression levels. These genes were involved in inflammatory and stress responses, signal transduction, and energy metabolism. However, in the physically active animals, the number of genes showing expression changes with age was significantly lower (n = 62) than in the sedentary animals. Moreover, physically active animals showed smaller fold changes as compared to the sedentary animals in 32 of the 42 genes that were common to both groups

(Bronikowski et al., 2003). These results suggest that regular exercise may retard several aging-related changes in cardiac gene expression.

Needless to say, similar data in humans are still missing. Furthermore, there are no data available on the genotype–physical activity interactions on health-related fitness in elderly subjects. Both of these topics are highly relevant areas for future research.

Summary and Conclusions

The past decade has witnessed remarkable progress in human genetics. The availability of the almost complete DNA sequence of the human genome has changed our ability to study the genetic basis of complex multifactorial traits and to develop novel treatments for several chronic diseases. The recent advances in molecular genetics are starting to have an impact on exercise physiology. Although the research on molecular genetics of physical performance and health-related fitness is still in its infancy, we recognize that understanding the effects of DNA sequence variation on interindividual differences in responsiveness to acute exercise and regular exercise training holds great promise. Such data would help to improve both athletic training and the use of physical activity in the prevention and treatment of chronic diseases. Furthermore, the availability of powerful methods such as microarray technology to measure gene expression, targeting specific genes in knockout and transgenic animal models, will greatly add to our research capability in the investigation of basic and applied exercise physiology issues across the life span.

Chapter 2

Physical Activity and Older Adults

Impact on Physical Frailty and Disability

Miriam E. Nelson, PhD, FACSM
Rebecca Seguin, MS, CSCS

There are currently 33 million Americans aged 65 and older living in the United States, and over the next century, that number will more than double, with the greatest increase occurring among individuals 85 and older (U.S. Department of Health and Human Services, 2000). The implications of extended years of life often involve increased incidence of chronic disease as well as the development of functional limitations and disability. Despite the recent reduced prevalence of disability, 7 million older adults still are chronically disabled (Manton & Gu, 2001). Physiologically, there is a loss of muscle mass and strength as age increases. Observational studies indicate that approximately 1% of muscle mass is lost per year after the fourth decade of life (Baumgartner et al., 1998; Janssen et al., 2000). This age-related loss of muscle mass is known as sarcopenia (Rosenberg, 1989). Sarcopenia can be caused or exacerbated by (or both caused and exacerbated by) certain medical conditions, yet it remains an issue of concern for the aging population regardless of chronic disease presence. Conversely, the loss of muscle mass and strength with age can increase an individual's risk for developing certain chronic conditions such as osteoporosis (Roubenoff & Hughes, 2000).

This material is based upon work supported by the U.S. Department of Agriculture under agreement #58-1950-9-001. Any opinions, findings, conclusions, or recommendations expressed in this publication are those of the author(s) and do not necessarily reflect the view of the U.S. Department of Agriculture.

Often concurrent with increased physical impairment is decreased ability to perform functional tasks such as climbing stairs, standing up from a chair, and doing basic household chores, all tasks that require a threshold of muscular strength. In some instances, it is underlying chronic disease presence that causes physical impairments; in other cases, it is simply age-related decrements, such as the loss of muscle mass and strength, which lead to these functional limitations. In any case, this series of events can lead to disability, dependence, and increased morbidity and mortality for older adults (Roubenoff, 2000).

Physical inactivity is one of the strongest predictors of physical disability among older persons (Buchner et al., 1992; Carlson & Harshman, 1999). Several longitudinal observational studies reveal that regular physical exercise not only extends longevity, but also reduces the risk of physical disability in later life (LaCroix et al., 1993; Strawbridge et al., 1996; Ferrucci et al., 1999; Leveille et al., 1999; Wu, Leu, & Li, 1999). Research in the last several decades has shown that many of the age-related physiologic decrements older adults experience are not inevitable. The primary components of physical fitness are cardiorespiratory endurance, flexibility, body composition, power, balance/coordination, muscular endurance, and muscular strength. Each component has its role in preserving function, reducing risk for chronic health conditions, and averting disability with age (Foldvari et al., 2000; Chakravarthy, Joyner, & Booth, 2002). Specifically, studies have now shown that targeted exercise referred to as strength training (also known as weightlifting or progressive resistance training) has the power to combat weakness and frailty and their debilitating consequences (Roubenoff & Hughes, 2000). Functionally, strength training is an activity in which muscles move dynamically against weight (or other resistance), with small but consistent increases in the amount of weight being lifted over time.

Although it has yet to be determined whether the muscle mass, strength, and other benefits gained through strength training can actually prevent disability in older adults, scientific research as well as community implementation of strength training programs have shown that strength training is a safe and effective means by which to improve physical capabilities, reduce risk for falls, prevent functional limitations, and avert the development of certain chronic diseases or their symptoms in older adults.

Overview of the Health Benefits of Exercise and Older Adults

Research studies over the past two decades have produced compelling evidence supporting the feasibility and the benefits of targeted physical activity programs for older adults (Roubenoff & Hughes, 2000; Chakravarthy, Joyner, & Booth, 2002). In particular, the benefits of strength training include increased muscle and bone mass, muscle strength, flexibility, dynamic balance, self-confidence, and self-esteem. Exercise also helps reduce the symptoms of various chronic diseases such as arthritis, depression, type 2 diabetes, osteoporosis, sleep disorders, and heart disease (Nelson et al., 1994; Castaneda et al., 2002). In addition, research demonstrates that strength and balance training in older adults with functional limitations reduces falls (Campbell et al., 1997; Buchner et al., 1997; Campbell et al., 1999; Rubenstein et al., 2000).

Table 2.1 summarizes several research studies, from small, highly controlled interventions with just a few subjects to the more recent, larger, home-based trials on exercise. This literature review is not meant to be comprehensive; we have selected several studies to illustrate several important aspects of exercise in older adults. We limited this review to studies that involved adults aged 50 and older.

Initial Strength Training Research With Older Adults

Early research investigating the effects of strength training on muscle mass, strength, and function was limited and quite conservative in terms of the intensity of the exercise prescription. As positive results were observed, strength training protocols for older adults have become more progressive and, as a result, more potent in nature. This was demonstrated by the first high-intensity strength training study in older adults conducted by Frontera and colleagues at Tufts University in the mid-1980s (Frontera et al., 1988). Twelve healthy men between the ages of 60 and 72 years strength trained at 80% of their 1-repetition maximum (1RM) for knee extension and leg curl. Their prescription—eight repetitions, three sets, three days per week—was very similar to many of the protocols being used today in clinical, community, and home-based settings. These investigators were the first to observe significant increases in muscle mass ($\approx$10%) and strength ($\approx$150%) over the study period of 12 weeks (Frontera et al., 1988).

Frailty and Falls

The aforementioned study was conducted with healthy, community-dwelling men in their 60s and 70s. The findings from that study initiated novel investigations of the effects of strength training in frail institutionalized men and women in their 90s by Fiatarone and colleagues (1990). In an effort to examine how strength training could affect muscle weakness and impaired mobility in relation to functional limitations and risk for falls in this population, participants were enrolled in eight weeks of strength training. In addition to strength gains of 174% on average and a 9% increase in midthigh muscle area, researchers also observed a 48% improvement in tandem gait speed. This combined increase in muscle strength, mass, and walking speed has multiple implications for reducing physiologic and functional impairment, as well as decreasing risk for fall and

TABLE 2.1 Selective Review of Exercise Studies in Older Adults

Study	Population	Design/duration	Intervention protocol	Setting	Significant outcomes[a]
Frontera et al., 1988	Healthy men aged 60-72 yr; n = 12	Noncontrolled; 12 wk	Strength training; 3 sets; 8 reps/set; 80% 1RM[b]; knee extensors, knee flexors; 3 × wk	Laboratory based	• Strength ↑ >100% • MTMA[c] ↑ 11.4%
Fiatarone et al., 1990	Men and women aged 86-96 yr; n =10	Noncontrolled; 8 wk	Strength training; 3 sets; 8 reps/set; 80% 1RM; knee extensors, hip extensors; 3 × wk	Long-term care facility	• Strength ↑ >150% • MTMA ↑ 9% • Gait speed ↑ 48%
Fiatarone et al., 1994	Men and women aged 72-98 yr; n = 100	Randomized controlled trial (RCT); 10 wk; four groups: strength training, strength training plus nutrition beverage supplement, nutritional supplement, control group	Strength training; 3 sets; 8 reps/set; 80% 1RM; knee extensors, hip extensors; 3 × wk	Long-term care facility	• Strength ↑ 113% • Stair climbing power ↑ 28% • Gait speed ↑ 12%

Study	Population	Design	Intervention	Setting	Outcomes
Nelson et al., 1994	Women aged 50-70 yr; n = 40	RCT; 1 yr; strength training vs. control group	Strength training; 3 sets; 8 reps/set; 80% 1RM; 5 exercises; 2 × wk	Laboratory based	• Strength ↑ 35-76% • Bone density ↑ 1% • Balance ↑ 14% • TBMM[d] ↑ 1.2 kg
McCartney et al., 1995	Men and women aged 60-80 yr; n = 142	RCT; 42 wk; strength training vs. control group	Strength training; 3 sets; 10-12 reps/set; 80% 1RM; 7 exercises; 2 × wk	Laboratory based	• Strength ↑ 20-65% • MTMA ↑ 5.5% • Treadmill endurance ↑ 18%
Skelton et al., 1995	Women aged 75 yr and older; n = 52	RCT; 12 wk; strength training vs. control group	Strength training; 3 sets; 4-8 reps/set; progressive; 8 exercises of upper and lower body using body weight, rice bags, and elastic tubing	Small-group classes at laboratory 1 × wk + 2 × wk home based	• Strength ↑ 4-27% • Power ↑ 18% • No change in functional performance
Singh et al., 1997a, 1997b	Men and women aged 60-84 yr who had depression; n = 32	RCT; 10 wk; two groups: strength training and health education	Strength training; 3 sets; 8 reps/set; 80% 1RM; 5 exercises; 3 × wk	Laboratory based	• Strength ↑ 33% • Depression reduced • Sleep improved

(continued)

TABLE 2.1 *(continued)*

Study	Population	Design/duration	Intervention protocol	Setting	Significant outcomes[a]
Ettinger et al., 1997; Messier et al., 2000	Men and women aged ≥60 yr; n = 439	RCT; 18 months; three groups: aerobic, strength, and health education (HE)	Aerobic or strength training groups: ankle weights; 2 sets; 12 reps/set; 9 exercises	Facility-based classes for 3 months followed by 15 months home based	Relative to HE: • Pain ↓ 8% • Physical disability ↓ 8% • 6-min walk improved • Balance improved ns between aerobic and resistance training group
Campbell et al., 1997, 1999	Women aged ≥80 yr; n = 233	RCT; 2 yr with four initial home visits; control group received usual care and equal number of home visits	Strength and balance training; 2 sets; 8 reps/set; progressive; 3 × wk; 9 strength exercises (2 body weight; 7 with ankle weights); ~9 balance exercises	Home based	• ns increase in strength • Balance improved • Chair rise improved • Falls reduced [relative hazard = 0.69 (0.49-0.97)]
Taaffe et al., 1999	Men and women aged 65-79 yr; n = 46	RCT; 24 wk; 1, 2, or 3 days/wk, plus a control group	Strength training; 3 sets; 10 reps/set; 80% 1RM; 8 exercises (days/wk varied depending upon group assignment)	Laboratory based	• Strength ↑ 37-42% in all three exercise groups • Chair rise time ↓ in all three exercise groups

16

Study	Population	Design	Intervention	Setting	Results
Jette et al., 1999	Men and women ≥60 yr with functional limitations; n = 215	RCT; 6 months with two home visits; control group was placed on a waiting list for the intervention	35-min exercise video performing 11 exercises using elastic bands of varying thickness	Home based	• Strength ↑ 6-12% in lower body • Tandem gait improved 20% • Physical and overall disability ↓ 15-18%
Baker et al., 2001	Men and women aged 55-82 yr with osteoarthritis of the knee; n = 46	RCT; 16 wk with periodic home visits; control group received nutrition education program	Strength training; 2 sets; 12 reps/set; progressive; 3 × wk; 6 exercises (2 functional, 4 with ankle weights)	Home based	• Strength ↑ 71% • Pain ↓ 43% • Physical function ↑ 44% • Depression reduced
Thomas et al., 2002	Men and women aged ≥45 yr with osteoarthritis of the knee (average age 62 yr); n = 786	RCT; 2 yr with periodic home visits; four groups: strength training (ST) with periodic home visits, ST with PHV plus telephone contact, telephone contact alone, or no intervention	1 set; up to 20 reps; progressive; total exercise time of 20-30 min/day; multiple lower body exercises with elastic bands	Home based	• Strength improved • Pain ↓ 12% • Physical function improved • Knee stiffness reduced
Fielding et al., 2002	Women aged ≥65 yr (average age 73 yr); n = 30	RCT; 16 wk; high-velocity resistance training (HI) vs. low-velocity resistance training (LO)	trength training or power training; 3 sets; 8-10 reps/set; 3 × wk; knee extension and hip extension	Laboratory based	• Strength improved similarly in both groups (HI and LO) • Power improved significantly more in HI than LO

(continued)

TABLE 2.1 *(continued)*

Study	Population	Design/duration	Intervention protocol	Setting	Significant outcomes[a]
Castaneda et al., 2002	Hispanic men and women aged 58-82; N = 60	RCT; 16 wk; control group received usual care	Strength training; 3 sets; 8 reps/set; 70-80% 1RM; 5 exercises; 3 × wk	Laboratory based	• Strength improved • HgA1c[f] reduced • Abdominal fat ↓ • Systolic BP ↓ • Lean tissue ↑
Dunstan et al., 2002	Men and women 60-80 yr with type 2 diabetes; n = 36	RCT; 6 months progressive resistance training (PRT) plus moderate weight loss (MWL) vs. control plus MWL	Strength training; 3 sets; 8-10 reps/set; 75-85% 1RM; 9 exercises; 3 × wk	Laboratory based	• Strength improved • HgA1c[f] reduced • Lean tissue ↑ • Weight loss in both groups
Nelson et al., 2004	Men and women aged 70-92 yr who scored ≤10 on established population for epidemiologic studies of the elderly (EPESE)[e] at baseline; n = 70	RCT; 26 wk with periodic home visits; control group received nutrition education program	Strength and balance training; 2 sets; 8 reps/set; progressive; 3 × wk; 9 strength exercises (3 body weight; 3 ankle weights; 3 dumbbells); 2 balance exercises	Home based	• ns increase in strength • EPESE ↑ 26% • Balance ↑ 34%

[a]Compared to nonexercising control subjects; [b]1RM = 1-repetition maximum; [c]MTMA = midthigh muscle area; [d]TBMM = total body muscle mass; [e]EPESE = EPESE summary physical performance score, [f]HgA1c = hemoglobin A1c.

fractures (Fiatarone et al., 1990). Fiatarone and colleagues went on to conduct a similar but larger study in 100 men and women in the same long-term care facility and obtained similar results (Fiatarone et al., 1994).

The study by Campbell and colleagues in New Zealand demonstrated that strength and balance training at home (with encouragement to increase walking) in near-frail women (≥80 years) with minimal supervision is possible on a large scale (Campbell et al., 1997, 1999). After two years the women in the exercise group had experienced 31% fewer falls (37% fewer falls resulting in a moderate or severe injury). It is important to note that the older women in the exercise group were given instruction during just four home visits; otherwise the women performed all of the exercise sessions on their own at home.

It is clear that strength training can reduce or delay functional limitations as well as reduce physical impairment and falls. Another critical question within this area of research is the effect of strength training on the development and progression of age-related chronic diseases such as depression, arthritis, and type 2 diabetes—all of which, without intervention, can ultimately lead to disability.

Bone and Joint Health

Several studies have shown improvements in bone density with strength training in older adults (Nelson et al., 1994; Kohrt, Ehsani, & Birge, 1997; Cussler et al., 2003). In one such study, Nelson and colleagues conducted a longer-term strength training study in women aged 50 to 70 (Nelson et al., 1994). This randomized controlled study was designed to see if strength training over one year could reduce the risk of osteoporotic fractures by increasing bone mineral density in estrogen-deplete women. After one year of strength training two days per week, middle-aged women became stronger, gained muscle mass, improved dynamic balance, and had improvements in bone density over and above values in the control group. In a study by Cussler and others (that included other modes of exercise in the intervention), there was evidence of a linear relationship between bone mineral density change and total weight lifted over the year, indicating the importance of progression and intensity for improvements in bone (Cussler et al., 2003).

Laboratory-based interventions such as the study by Ettinger and colleagues have demonstrated beneficial (albeit modest) effects of both aerobic and strength training exercises on reducing the signs and symptoms of osteoarthritis in older adults (Ettinger et al., 1997; Sevick et al., 2000; Messier et al., 2000). In this 18-month trial, 439 adults aged 60 and older who had osteoarthritis of the knee were randomized to three groups: education control, aerobic exercise, or a resistance training program. Both exercise groups had a 3-month center-based intervention followed by a 15-month home-based intervention. After 18 months, the two exercise training groups had an 8% lower score on a physical disability questionnaire, had an 8% lower score on pain, showed improvements in balance, and walked 57 ft longer on the 6-min walk than the education control group. There were no differences between the two exercise groups. Interestingly, a cost-effectiveness evaluation demonstrated that the strength training intervention was more economically efficient than aerobic exercise in improving physical function—however, the magnitude of the difference between the two exercise groups was small (Sevick et al., 2000).

In 2001, Baker and colleagues at Tufts University published data regarding strength training in individuals with osteoarthritis (Baker et al., 2001). Forty-six community-dwelling individuals aged 55 and older with knee osteoarthritis were randomized to either a home-based strength training group or an attention control group (nutrition education) for four months. Patients in the exercise group experienced significant reductions in pain and improvements in muscular strength, functional performance, physical abilities, quality of life, and self-efficacy. Similarly, Thomas and colleagues in Nottingham, England, randomized 786 men and women with knee pain to either a self-reported home-based strength training group (with or without phone contact) or a control group (with or without phone contact) (Thomas et al., 2002). The strength training group experienced significant reductions in pain and stiffness and improvements in physical function compared to the control group at the 6-, 12-, and 24-month time points. These studies again illustrate that the home-based approach offers a feasible method for implementing strength training programs in older adults, even in those with functional limitations (Baker et al., 2001; Thomas et al., 2002).

Endurance

McCartney and colleagues completed a 42-week study of strength training in 60- to 80-year-old men and found not only improvements in strength and muscle mass but also improvements in treadmill endurance (McCartney et al., 1995). A study by Beniamini et al. showed similar improvements in treadmill endurance in subjects in a cardiac rehabilitation program when strength training was included in the exercise prescription (Beniamini et al., 1999). These two studies indicate that individuals with low aerobic capacity can realize important improvements in cardiovascular fitness with the inclusion of strength training.

Strength and Functional Performance

At the University of London, scientists studied the effects of 12 weeks of strength training on strength, power, and functional abilities in women aged 75 and older (Skelton et al., 1995). Fifty-two healthy women were randomized to either a three day per week strength training group or a control group, which received no intervention. Two of the three exercise sessions were completed at the study participants' homes while the other session was completed at the research center with supervision. Compared to control subjects, participants in the strength training group showed significant improvements (27%) in knee extensor muscle strength. Investigators also observed significant improvements in muscular power (18%). They did not observe an effect of the strength training program on functional tests, possibly because the study subjects were healthy and had no functional limitations at baseline (Skelton et al., 1995).

However, a recent investigation by Nelson and colleagues demonstrated significant improvements in physical function following six months of home-based strength and balance training in elderly subjects who had functional impairments prior to entry into the study (Nelson et al., 2004). Seventy independently living men and women aged 70 and older were randomized to either a control group or an exercise group, the latter performing strength and balance training three times per week in their own home. The exercise group experienced a 26% improvement in physical function. These findings highlight the importance of targeted exercise programs for both preventing function decline and improving physical abilities with age. Furthermore, this study demonstrates that it is possible for near-frail elders to strength train at home with minimal supervision and to have meaningful improvements.

A Variety of Strength Training Prescriptions

There is also the very important question of frequency of training. Currently the American College of Sports Medicine recommends strength training for musculoskeletal fitness two to three days per week (American College of Sports Medicine, 1998). The question arises whether physiologic improvements can result with just one day per week of strength training. Another six-month intervention addressed the effects of one, two, or three days per week of strength training in 46 community-dwelling elders compared to a control group (Taaffe et al., 1999). All those participating in strength training experienced significant increases in muscular strength and improvement in chair stand time compared to the control group, and there was no difference among the three exercise groups. This study demonstrated that as little as one day per week of strength training improved strength and physical function. Questions remain as to whether one day a week is enough to improve health outcomes such as reducing the pain associated with arthritis. Data from this and similar studies are important when one considers a safe, effective, and reasonable exercise prescription for healthy older adults in terms of preventing functional limitations and physical impairment (Taaffe et al., 1999).

The Potential of Power Training

As the area of strength training investigation progresses, some scientists are trying to identify the best models for implementation and dissemination, while others continue their work in determining what exercise prescription will confer the most benefits for older adults. New research from Fielding and colleagues provides insight into how training for power (defined as the combination of strength and speed) in this population may effectively target functional performance (Fielding et al., 2002). These authors randomized 30 women with self-reported disability to either a high-velocity strength training program or a more traditional low-velocity strength training program. Subjects completed three lower body exercises at 70% 1RM three times a week for 16 weeks. Those in the high-velocity strength training group experienced significantly greater improvements in skeletal muscle power, which because of the combination of speed and strength may be an important factor in warding off functional impairment and therefore an important consideration in designing exercise programs for older adults (Foldvari et al., 2000).

Conclusion

The effect of exercise training on physical and functional status in older adults is a relatively new field of investigation. We are only just beginning to understand the full potential of various modes of exercise (i.e., strength training and balance training) in preventing and controlling various disease states as well as the implications for reducing functional impairments and disability. Many age-related physiologic changes occur hormonally, neurologically, metabolically, and behaviorally that contribute to sarcopenia. Exercise can have a positive impact on each of these physiologic domains and is therefore both a viable and a potent adjunct to any physical activity prescription for older adults.

What is not fully understood is the impact of exercise on preventing or delaying the development of disability with age. It is important that in further examining this issue, investigators are discriminating about the assessment tools selected so that both specificity and sensitivity to change are accurately measured and interpreted. Future investigations should also focus attention on identifying the specific mechanisms that catalyze physiologic changes as a result of diverse exercise training interventions.

Chapter 3

Exercise Dose-Response Effects in Older Adults

Roy J. Shephard

Regular physical activity carries many important benefits not only for younger people (Bouchard, Shephard, & Stephens, 1994; Bouchard et al., 1990), but also for senior citizens (Shephard, 1997a). However, it is not easy to motivate people to become more active at any age (Shephard, 1994b), and an understanding of the dose of physical activity needed to maximize health benefit is central to such efforts (Shephard, 1997c). Unfortunately, there is disagreement on the message that we should proclaim, and this weakens the impact of the advice that is offered (Shephard, 2002).

Relative Versus Absolute Intensity of Effort

Should we recommend exercise at a certain relative intensity, for example 60% of the individual's maximal oxygen intake? Or should we prescribe an *absolute* energy expenditure such as a minimum *intensity* of 6 METs, or a minimum *volume* such as 8 MJ/week (Shephard, 2001)? This is not a new debate, as we can see from a review published in 1965:

> Some workers, basing their views on personal experience of illness or on the one experimental study of bed rest[52], have contended that the ordinary routine of daily life produces considerable training. Others[14, 32] have stated categorically that mild exercise, such as golf and bowling is not enough; a pulse rate of 140/min (60% of the maximum possible increase over the resting value[32, 31]) or 150/min[34] is essential for training. (Shephard, 1965, p. 533)

Relative and absolute intensity of effort are plainly interrelated, and indeed may be very similar to each other in a population that is homogeneous with respect to age, gender, and fitness. But the choice between relative and absolute intensity becomes critical when we try to extrapolate findings on young adults to the older population, since a fixed absolute intensity of effort demands a much greater relative effort of the average senior than it does of a 20-year-old (Bouchard et al., 1990). Plainly, a training heart rate of 140 beats/min would represent supramaximal aerobic exercise for a person aged 65, and would be impossible for most people in middle- or old-old age (Shephard, 1997a). We can overcome the difficulty in classifying the intensity of aerobic exercise if we express the oxygen consumption required for a given activity as a fraction of the individual's oxygen consumption reserve (Howley, 2001).

Relative intensity is usually calculated with respect to the person's peak oxygen intake, as measured on the treadmill. But because muscle strength declines with aging, there are substantial differences of peak oxygen intake between cycle ergometer and treadmill tests (Shephard, 1994a). Such differences are exacerbated if only a small muscle mass is activated (for example, in arm ergometry); intensities must thus be expressed relative to the peak oxygen intake a person can develop when performing a specific mode of exercise.

Experimental Evidence

Investigators in an early experimental study trained a group of 39 young sedentary men at three different intensities, 39%, 75%, or 96% of their individual maximal aerobic power, using sessions of 5-, 10-, or 20-min duration with a frequency of one, three, or five sessions per week (Shephard, 1968). A stepwise multiple regression analysis showed that gains in maximal oxygen intake were related to the individual's initial maximal aerobic power (expressed in ml · kg^{-1} · min^{-1}), to the intensity of training (also expressed in ml · kg^{-1} · min^{-1}), and to the number of training sessions per week. Some conditioning was seen even at the lowest dose of training, but the size of the response was influenced most strongly by the intensity of training relative to the individual's personal level of aerobic fitness.

No one has yet carried out a comparable experiment on senior citizens. It is difficult to assess gains of maximal oxygen intake in elderly people. A first stress test may be halted prematurely; but after training, both the subject and the investigator are more confident in pursuing a true maximal effort. A simple comparison of initial and final data may thus overestimate the extent of training responses. The synthesis of protein proceeds more slowly with aging (Welle et al., 1993), but seniors can enhance the function of both cardiovascular and musculoskeletal systems. The response may be limited by the ability of aging tissues to withstand frequent and intensive training without injury; but if training is undertaken at the same relative percentage of maximal effort as in a young person, we may anticipate a similar influence from the intensity, frequency, and duration of exercise sessions.

Some Problems of Research Design

It seems logical that for any given dose of exercise, training responses will be greatest in those who are unfit; but unless appropriate precautions are taken, at least a part of this apparent effect reflects a regression of data toward their mean values (Shephard, 2003). Statisticians recommend adoption of a true experimental design, with randomized assignment of subjects to experimental and control programs. But many people are unwilling to accept exercise randomization. Those who wish to exercise are reluctant to refrain from exercising, and others with a sedentary disposition quickly drop out of an exercise cohort. Studies that achieve a successful experimental design are thus based on a small number of subjects, often young university students, and their responses may not reflect what will occur in the general elderly population.

Epidemiological studies usually have a larger sample size but may also suffer from a lack of generalizability. For example, much has been learned from the sophisticated analyses of Steve Blair and his colleagues (Blair, Cheng, & Holder, 2001; Blair & Connelly, 1996; Blair et al., 1995, 1989), but some critics have argued that the high socioeconomic status of their subjects limits generalizations to wider populations in the United States or elsewhere. Because of practical constraints, many investigators make inferences from nonrandomized observational studies—cross-sectional comparisons of active versus sedentary individuals, or case-control studies. Plainly, problems then arise from associations between a physically active lifestyle and other positive prognostic features such as a young biological age, a high socioeconomic status, and a favorable lifestyle (Shephard & Bouchard, 1996)—for example, the greater longevity of participants in endurance sport may simply reflect the fact that the majority of them are nonsmokers (Shephard, Kavanagh, & Mertens, 1995).

A further important issue is the choice of outcome measure. Ideally, patterns of physical activity should be related to quality-adjusted life expectancy (Shephard, 1996), using a prospective experimental design. But in practice, granting agencies and investigators are rarely willing to wait until a cohort of experimental subjects has died or developed a medical condition that impairs their quality of life. So, conclusions are based on the change in some surrogate marker of health outcome. Suitable surrogates include not only maximal oxygen intake but also exercise-induced changes in systemic blood pressure (as a marker of stroke risks), serum lipids (as a marker of cardiac health), blood glucose regulation (indicating the risk of maturity-onset diabetes mellitus), and various indices of immune function (Shephard, 1997b). Often, the health educator must choose between competing health indices, assessing their respective impacts on the individual's quality-adjusted life expectancy and selecting the most important outcome (Shephard & Bouchard, 1996). This point can be illustrated by a cross-sectional study of some 350 people conducted in Quebec City (Shephard & Bouchard, 1995). Whereas gains in cardiovascular variables were closely linked to perceptions of participation in vigorous physical activity, the control of body fat and other metabolic problems was more closely associated with reports of frequent, moderate bouts of physical activity.

Is Aerobic Fitness Enough?

An increase of aerobic fitness reduces the risk of cardiovascular and all-cause mortality (Shephard, 1994a, 1997a) and maintains functional independence and thus the quality of life as aging progresses (Shephard, 1996). Many early observers seem to have assumed that a senior's health needs would be satisfied by a dose of exercise that optimized such surrogate markers of cardiorespiratory health as maximal aerobic power. However, gains in maximal oxygen intake do not necessarily reflect success in enhancing other aspects of long-term health (Bouchard, Shephard, & Stephens, 1994; Bouchard et al., 1990). Other important issues for an elderly population include metabolic health (the control of obesity, an appropriate regulation of blood glucose, the prevention of maturity-onset diabetes mellitus, and optimization of the lipid profile), correction of osteoporosis, enhancement of a waning immune function, reduction in cancer risks, and mental health. Such issues are not necessarily addressed by a dose of exercise that increases aerobic fitness (Shephard, 1997a).

Moderate or Intense Physical Activity?
A Public Policy Debate

The issue of whether to advocate moderate or intense physical activity is far from resolved. It is difficult to adopt an experimental approach. If an activity such as walking or attendance at an exercise class is assigned experimentally, it may provoke a reduction of physical activity during the remainder of the day, sometimes to the point that there is no net increase in weekly energy expenditures. Retrospective analyses are complicated by difficulty in recalling activity patterns. For example, the association of cancer prevention with *concordant* reports of an active lifestyle as younger and elderly adults may indicate the importance of long-sustained active behavior, but it may also reflect the greater accuracy obtained by duplicate assessments of physical activity (Lee et al., 1997). Further, vigorous activity is usually recalled more precisely than light or moderate effort.

In epidemiological studies, activity patterns are often associated with overall health attitudes (Shephard & Bouchard, 1996). Thus, the lean nonsmoker tends to be physically active. Health outcomes can be adjusted statistically for differences in other facets of lifestyle; but if we make such adjustments, previously observed linkages between physical activity and health may disappear.

Because many older people are not too enthusiastic about *maximizing* their functional capacity, debate has shifted to a discussion of the *minimum* level of physical activity needed to achieve the desired health-related outcomes. Experts now encourage "active living"—the deliberate incorporation of physical activity into everyday life. For example, using walking or cycling as a means of transport and adopting hand rather than power tools around the home (Health Canada, 1999). Most daily activities require only a moderate intensity of effort and are thus of greatest benefit to those in whom a moderate *absolute* intensity of effort also develops a substantial *relative* effort (Shephard, 1997c), for example, sedentary elderly and obese individuals.

Most field observers lack the information necessary to estimate relative intensities of effort, and many papers thus have asked whether health outcomes are enhanced by some absolute energy expenditure, measured in METs, METs/week, kJ/min, or MJ/week. One well-publicized study claimed that cardiovascular mortality was reduced if absolute exercise energy expenditures exceeded a minimum of 2 MJ/week (Paffenbarger et al., 1986). However, this conclusion is challenged by (a) cumulative errors of up to 2 MJ/week in the assumed energy cost of one of the principal activities (stair climbing) and (b) a failure to realize that, depending on the duration of activity, expenditures between 1 and 2 MJ/week must be imputed to normal resting metabolism during the minutes when exercise is performed (Shephard, 1999).

Consensus Recommendations

Given difficulties in organizing effective experimental and epidemiological trials, some credence may be attached to the findings of expert groups. At least 23 national and international consensus reports have appeared over the past 10 years (Shephard, 2001), although the Canadian Society of Exercise Physiology/ Health Canada document is unique in making a specific recommendation for senior citizens (Health Canada, 1999). A few groups have looked at specific outcomes such as the prevention of hypertension, obesity, or coronary disease, but most have taken a global approach to health. The emphasis has been on avoiding premature death, rather than on the more important issue of maximizing quality-adjusted life expectancy. Further, there has been little attempt to evaluate the relative importance of competing health outcomes. Often, reports have recommended a minimum *duration* of physical activity for any given relative intensity of effort. Thus, de facto, a minimum fitness-related *volume* of physical activity has been specified. However, most groups have failed to discuss the importance of relative versus absolute intensity of physical activity.

Almost all expert groups have concluded that the appropriate *minimal* recommendation is light to moderate effort (40-60% of $\dot{V}O_2max$ or 40-50% of $\dot{V}O_2R$). However, the apparent unanimity of opinion must be viewed with some skepticism, given the overlap in membership among various consensus groups. Some reports have suggested additional health benefits if people progress to a higher intensity of effort (up to 85% of $\dot{V}O_2max$ or $\dot{V}O_2maxR$), but many recent statements argue that "active living" provides an adequate minimum of physical activity.

Systematic Review of Dose-Response Issues

The Hockley Valley symposium undertook a systematic review of dose-response issues (Kesaniemi et al., 2001; Shephard, 2001), and these publications should be consulted for a more detailed bibliography. Information on absolute versus relative intensity of exercise (Shephard, 2001) was sought in Medline, Sport Discus, 13 major reviews, and the author's personal files. During the period 1991 through December 1999, 1,666 papers discussed *exercise + intensity,* and 770 *exercise intensity;* but a combination of the keywords *intensity + exercise + health* yielded only 24 citations. Likewise, the keyword *absolute exercise intensity* yielded 7,528 citations, but when this was combined with *health,* the total dropped to 52. *Relative intensity* yielded 4,441 papers, but this was reduced to 5 citations with addition of the term *health.* Sport Discus was reviewed for the period January 1995 to September 1999. Given its exercise orientation, the search of this database was limited to the keywords *exercise prescription, intensity,* and *health;* this yielded 130 hits.

Many of the papers that were identified merely asserted that a certain relative or absolute energy expenditure had a beneficial influence on health. The final analysis was limited to 166 articles whose authors had sought evidence of a threshold of relative or absolute intensity of effort. Unfortunately, the importance of relative versus absolute energy expenditure was often obscured even in this sample, because an increase of intensity was allowed to increase the absolute energy expended. Most reports were based on older adults, probably because this increased the likelihood of determining health outcomes, but specific effects of aging were rarely considered.

All-Cause and Cardiovascular Mortality

Of individual health outcomes, all-cause and cardiovascular mortality received the greatest attention. Many early studies were cross-sectional occupational comparisons; these are compromised by a substantial fitness-related selection into and out of heavy employment, and at best one can distinguish light from hard physical work. Twenty-nine large-scale longitudinal surveys have related simple questionnaire estimates of leisure activity to cardiac outcomes. In 16 of these reports, the type, frequency, intensity, and duration of physical activity were combined to yield semiquantitative estimates of the volume of weekly energy expenditures, sometimes described erroneously as the "intensity" of effort. By way of example, Harvard Alumni data suggest that total and cardiovascular mortality are reduced progressively over the total weekly energy expenditure range 2.1 to 8.4 MJ/week (Paffenbarger et al., 1986). Fourteen studies looked specifically at differences in outcome between individuals who claimed a vigorous intensity of physical activity versus those who did not. In many of these studies, the threshold for benefit was an intensity of around 6 METs (Lee & Paffenbarger, 1996). However, three comparisons between moderate and hard physical activity (Bijnen, Caspersen, & Mosterd, 1995; Manson et al., 1999; Rosengren & Wilhelmsen, 1997) showed health benefits at lower intensities of effort. Furthermore, in one report it was observed that hypertensive subjects were at increased risk during intensive effort.

Unfortunately, few studies have examined the absolute intensity range recommended by many expert groups (that is, 4-6 METs). Moreover, the threshold for health benefit usually has been reported in absolute units, so that the relative intensity of effort has been unknown. Additionally, few investigators have differentiated gradations of exercise intensity from associated changes in the volume of activity. Some analyses of the Harvard Alumni study suggested added benefit from intensity at any given absolute volume of physical activity (Lee & Paffenbarger, 1997). Other investigators noted an effect of intensity independent of volume or an effect of duration independent of intensity. A few studies have looked at other markers of cardiovascular risk. A total weekly energy expenditure of 8.4 MJ seems necessary to induce radiographic regression of atherosclerotic lesions. Several small crossover laboratory trials have also looked at acute exercise-induced changes in clotting mechanisms. With one exception, these suggest that benefit is much greater with hard than with moderate intensity physical activity.

The Hockley Valley symposium (Kesaniemi et al., 2001) rated studies in terms of the *quality* of evidence presented. The highest weight, grade A, was assigned to large-scale randomized trials. Most of the evidence on cardiovascular disease fell into category C, which comprises uncontrolled or nonrandomized trials and cross-sectional and prospective observational studies. A substantial proportion of reports noted that a threshold absolute intensity of physical activity of around 6 METs was needed for benefit. Usually, this threshold was not expressed relative to the individual's maximal aerobic power, so we still do not know the minimum relative intensity of physical activity that is needed. Moreover, most studies have failed to distinguish the effects of intensity independently of associated changes in the volume of physical activity performed.

Stroke

Many studies of stroke have focused on the intermediate marker of blood pressure change. Three small-scale crossover trials agreed that the acute reduction of blood pressure following a single bout of exercise was independent of relative intensity, but a single large cross-sectional analysis found that the velocity of running (and thus the absolute intensity of effort) had a much greater impact than the total volume of activity as indicated by the distance covered. Nine of thirteen chronic studies involved small samples of randomly assigned subjects. Three of the remaining four reports were based on large-scale nonrandomized longitudinal studies. The results are conflicting. Four groups of investigators (Hagberg et al., 1989; Kingwell & Jennings, 1993; Matsusaki et al., 1992; Rogers et al., 1996) found a U-shaped relationship, with an adverse effect on blood pressures at high intensities of effort. Since the total volume of activity was not controlled, adoption of too high an intensity may have reduced the total volume of effort, and thus the impact on associated risk factors such as obesity and blood lipid profile. In contrast, four research groups (Braith et al., 1994; Dunn, Andersen, & Jakicic, 1998; Roman et al., 1981; Tashiro et al., 1993) noted no effect of intensity. Three other reports showed an *enhanced* response with more intense effort (Duncan, Gordon, & Scott, 1991; Folsom et al., 1990; Haapanen et al., 1998). Paffenbarger's cross-sectional study suggested that the threshold for a lasting reduction of blood pressure was an energy expenditure of 8.4 MJ/week, but he did not examine the importance of relative versus absolute intensity of effort in reaching the threshold (Paffenbarger et al., 1983).

Five cross-sectional analyses and a case-control study have looked at the influence of physical activity on the risk of stroke, although failing to differentiate between hemorrhagic and ischemic strokes. Physical inactivity apparently increased the risk of hemorrhagic stroke, but in ischemic stroke this was true only of smokers. Two reports (Linsted, Tonstad, & Kuzma, 1991; Menotti & Seccareccia, 1985) suggested a U-shaped relationship between intensity and the risk of stroke, with little benefit at intensities >70% of $\dot{V}O_2$max. Kiely and associates found no effect of intensity (Kiely et al., 1994), and other reports (Herman et al., 1983; Paffenbarger et al., 1983; Sacco et al., 1998; Wannamethee & Shaper, 1992) indicated a larger response with greater effort, although in none of these studies was the total volume of exercise controlled.

In summary, category B evidence from small-scale randomized trials suggests that the acute, exercise-induced reduction of blood pressure is independent of the relative intensity of exercise, but the long-term benefit may be greater for moderate than for hard effort. The optimal intensity to prevent stroke remains unclear.

Obesity

Eight small-scale crossover trials and one small randomized study examined the *acute* effect of exercise upon resting metabolism. Five of these trials controlled the relative intensity of effort for the *total* volume of activity performed. Five trials showed an intensity-related increase in either the *amount* (Broeder et al., 1991; Short, Wiest, & Sedlock, 1996; Treuth, Hunter, & Williams, 1996) or the *duration* (Bahr & Sejersted, 1991; Smith & McNaughton, 1993) of the excess postexercise oxygen consumption. Three other reports (one with control of total volume) showed no effect of intensity (Goben, Sforzo, & Frye, 1992; Sedlock, 1991; Segal et al., 1992), and Chad (Chad & Quigley, 1991) noted that at 3 hr postexercise the stimulation of metabolism was greater from activity at 50% of $\dot{V}O_2$max than from a bout at 70% $\dot{V}O_2$max.

In terms of fat loss, the total volume of physical activity and thus the energy expenditure might seem more important than the intensity of effort. Nevertheless, intense exercise could facilitate fat loss by increasing resting energy expenditure or lean body mass (and thus the ability to undertake a prolonged bout of endurance exercise). Eight small-scale randomized trials and four larger cross-sectional studies disagree as to whether the response is unchanged or increased by an increase in the relative intensity of exercise, although it may be important that the *total* volume of physical activity was controlled in one of two studies that reported *no* effect from the intensity of activity. Two reports compared the fat loss from structured and lifestyle activities. The intensity of activity was less in the lifestyle program. Fat loss was *similar* for the two groups; but in one of these comparisons, the lifestyle group lost more lean body mass than the participants in structured activities. A high intensity of effort seems helpful in conserving lean tissue during fat loss, an issue that is particularly critical in elderly people with weak muscles. Two further trials showed no significant changes in overall body mass relative to control at any of two or three intensities of effort. One study indicated reductions in skinfold thicknesses with both resistance and aerobic exercise.

Lipid Metabolism

One small randomized trial and six small-scale crossover trials have examined the acute impact of physical activity upon lipid metabolism. Five of the crossover trials controlled for the *total* volume of physical activity performed. Postprandial lipemia was unaffected by exercise intensity. One trial showed no acute changes in the lipid profile; but in the two remaining studies, there was a significant effect of relative intensity. Friedlander and associates found a greater effect on plasma free fatty acid kinetics with a high intensity of activity in men, but not in women (Friedlander et al., 1998).

Chronic changes in lipid profile have been examined in eight small-scale randomized longitudinal trials and seven cross-sectional studies. Five of these studies controlled for the total distance walked. Four of the randomized trials showed little change in lipid profile relative to control at any of two or three relative intensities of effort. Two other reports showed greater benefit from moderate- than from hard-intensity programs, and in the remaining two studies there was no effect of relative intensity. The majority of cross-sectional studies also showed an effect of exercise *volume*. Williams (1998) noted specifically that the effect on lipid profile was six times larger for running *distance* than for running *velocity*. Thus, it seems that volume of activity is the critical determinant of lipid optimization.

Glucose Regulation and Diabetes Mellitus

Seven small crossover trials, one randomized trial, and two cross-sectional studies have looked at exercise intensity versus glucose utilization, insulin sensitivity, or both. Three studies controlled for the total volume of activity.

Three of four trials showed a greater increase of glucose uptake at the higher intensity of effort, the exception being a study of non-insulin-dependent cases (Braun, Zimmerman, & Kretchmer, 1995). Two of three trials also yielded an improvement of insulin sensitivity only at the harder intensity of effort. Five large-scale cross-sectional studies explored the risk of developing diabetes mellitus in relation to habitual physical activity. Two studies pointed to a need for activity of sufficient intensity to induce a sweat, and two others found progressive protection from an increased total *volume* of energy expenditure; but no investigator has yet explored the issue of absolute versus relative intensity.

In summary, investigators of metabolic benefits have sometimes failed to distinguish between acute and chronic physical activity. Normalization of lipid profile depends on the total *volume* rather than the intensity of physical activity. However, the excess postexercise oxygen consumption and thus fat loss seem to depend on the relative intensity of effort, and activity programs of sufficient intensity to enhance lean body mass may increase the ability of the frail elderly to exercise and thus improve their metabolic regulation.

Osteoporosis

The extent of weight bearing and the application of resistive force are probably more important to the prevention of osteoporosis than either the relative or the absolute intensity of physical activity. Available information includes data from two small randomized controlled trials and two larger cross-sectional studies, all performed on postmenopausal women. The cross-sectional observations seem to suggest benefit from the volume of activity and from intensity if uncontrolled for volume. However, Pruitt and associates found similar effects from contractions at 40% and 80% of 1-repetition maximum if the total volume of activity was controlled (Pruitt, Taaffe, & Marcus, 1995). Some recent observations suggest also that the intervals between mechanical impulses may be important to enhancement of bone density.

Immune Function

Most studies of exercise and the immune system have looked at acute responses. Six small-scale studies examined the effects of various intensities of effort; three of these studies controlled for the total volume of physical activity. One report indicated no change in serum immunoglobulin A at any intensity of effort, but four of the five remaining studies showed a transient immunosuppression only at a high relative intensity of effort (75-80% of $\dot{V}O_2max$). There is some cross-sectional evidence that regular distance running protects against the deterioration of immune function seen in the elderly (Shephard, 1997b).

Cancers

A decrease in the risk of certain cancers is an important benefit from sustaining regular physical activity over many years. Seven large-scale nonrandomized longitudinal studies and a meta-analysis explored the effects of exercise volume, intensity, or both on all-cancer death rates. A meta-analysis showed greater benefit from a large than from a moderate dose of physical activity. However, individual reports are inconsistent, and in most cases the volume of activity has not been distinguished clearly from intensity (Shephard & Futcher, 1997).

Research has focused particularly on colon cancers. Here, data are drawn from six nonrandomized longitudinal studies, three case-control studies, and a meta-analysis. Again, the issue of relative intensity has not been resolved. The most convincing single study used the case-control technique to show a much greater protection from hard- than from light-intensity activity, but rather similar effects from small or moderate volumes and a large volume of physical activity. In the meta-analysis, protection against colon cancers was greater for light than for moderate activity.

Mental Health

Fourteen studies have looked at the effects of exercise intensity upon mental health, eight considering acute and six chronic responses. Most are based on relatively small samples of subjects; moreover, with two exceptions, the subjects have had only minor (nonclinical) disturbances of affect. Four of the acute trials showed a larger benefit at high relative intensities of effort; but, with one exception, the effect of a harder intensity was not distinguished from that of an increased total *volume* of physical activity. In terms of controlling long-term anxiety, the evidence suggests (with one exception) that a moderate relative intensity of exercise may be more effective than a hard intensity.

Adverse Outcomes

Potential adverse effects of excessive effort include musculoskeletal injuries, cardiac incidents, immune suppression, and overtraining. Such outcomes could theoretically reduce the *net* benefit to the population. One cross-sectional comparison showed that occasional activity at an intensity of over 6 METs (less than once per week) was associated with a 107-fold increase in the risk of a myocardial infarction in the first hour postexercise, whereas there was only a 2.4-fold increase of risk in those who exercised at the same intensity five or more times per week (Mittelman et al., 1993). Fortunately, the elderly are naturally cautious. The risk of such complications seems greater in middle than in old age, and few seniors are likely to reach the intensities of effort associated with adverse outcomes.

Research Priorities

What priorities should we set for future research? Plainly, more precise methods are needed to assess both relative and absolute energy expenditures in epidemiological work. Attention should be focused on gender- and fitness-related differences in physical activity requirements at various points in the life course. Future studies must equate absolute energy expenditures between several differing relative intensities of effort. Given that differing patterns of effort favor different health outcomes, we must also assess which types of outcome contribute most to a person's quality-adjusted life expectancy.

We need to explore whether there are multiple mechanisms of health benefit, with one mechanism induced by the absolute intensity or volume of effort and another by relative intensity. We must distinguish possible differences in intensity requirements between the prevention and the treatment of disease. Finally, we must clarify optimal as well as minimal intensity requirements and set ceilings of relative and absolute intensity. We will then be in a much better position to make public health policy recommendations for both young and elderly people.

Chapter 4

Control and Regulation of Movement in Elderly Adults

Caroline J. Ketcham
George E. Stelmach

The control of movement is a complex interaction of cognitive and sensorimotor systems. Researchers in movement science aim to understand how an action is produced and what mechanisms are involved in the control and regulation of movement. In recent years, advances in recording techniques have introduced new and informative ways to examine motor performance. Such advances have also introduced new and different measurement and evaluation issues that are just beginning to be debated. As new recording and analysis methods are scrutinized, it is likely that better insights into how the brain controls and regulates movement will emerge. This new information will also provide a better understanding of why the control and coordination of movement decline with advanced age.

Researchers over the years have conducted numerous studies to compare and contrast young and elderly age groups. These studies have used a variety of tasks and methods of analysis to document declines in older adults in both the central and peripheral nervous system that result in an array of behavioral decrements (Salthouse, 1985; Welford, 1977, 1984; Ketcham & Stelmach, 2002). Moreover, it is well known that as adults age, the execution of movement becomes slow and more variable, and there is emerging evidence that the microstructure of the movement also changes. As will be apparent in this presentation, research on the elderly has moved beyond surface descriptions of movement deficits into analyses that attempt to address motor performance differences at a mechanistic level (Cooke, Brown, & Cunningham, 1989; Darling, Cooke, & Brown, 1989; Goggin & Meeuswen, 1992; Ketcham et al., 2002; Pratt, Chasteen, & Abrams, 1994; Walker, Philbin, & Fisk, 1997; Seidler-Dobrin & Stelmach, 1998). This shift in focus is promising but involves numerous measurement and evaluation issues.

In the studies reviewed here, older adults are classified as over 60 years of age and are compared to young adults who are typically between 18 and 30 years of age. Results reported are means derived from cross-sectional comparisons, which attempt to document the characteristics of performance between individuals in different age groups. Thus, these comparisons provide only a snapshot into how age influences motor performance. While longitudinal research that documents how individuals change over decades provides the most insights into the aging process, this type of research is seldom conducted because of the long time periods involved (Fozard et al., 1994; Spirduso, 1995).

Before we begin to describe how age influences performance, it is important to recognize that when movements are performed there are well-documented speed–accuracy trade-offs in young and elderly subjects that often make comparisons between age groups difficult. It is well known that the elderly typically move slowly in an attempt to preserve movement accuracy (Salthouse, 1985; Salthouse & Somberg, 1982; Welford, 1977; Morgan et al., 1994; Ketcham et al., 2002; Goggin & Meeuswen, 1992). Thus in research that demonstrates age effects between age groups, it is important to recognize how a speed–accuracy trade-off can influence interpretations. Each time a difference between young and elderly populations is obtained in the time domain, it is necessary to ask if there are differences in accuracy (error rate) levels between the age groups. Unless one knows the speed–accuracy profiles of each, it is not possible to completely ascertain whether an experiment captures group differences in speed, accuracy, or some combination of both.

Figure 4.1 presents a hypothetical speed–accuracy relationship. As can be seen, when movement accuracy is plotted as a function of move-

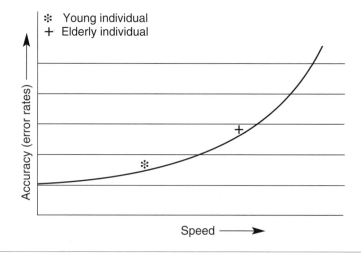

FIGURE 4.1   Performance accuracy plotted as a function of performance speed for a young and an elderly subject. Depending on the current strategy, an individual can attempt to regulate speed, accuracy, or both.

ment speed, accuracy decreases as speed increases. On the basis of their pre-ferred speed–accuracy strategies, individuals perform at a speed level at which they are comfortable. It is important to note that each individual has his or her own speed–accuracy curve and can move up and down the curve by regulating speed (Salthouse, 1985; Salthouse & Somberg, 1982; Welford, 1977; Morgan et al., 1994; Ketcham et al., 2002; Goggin & Meeuswen, 1992). It should be apparent that it is necessary to establish speed–accuracy relationships before making age-group comparisons in order to account for strategy differences between groups.

In this presentation we begin with movement execution changes that occur in older adults. Initially movement characteristics (duration, kinematic pro-files, and force production) are examined in tasks that primarily utilize upper extremity goal-oriented movements. Here we focus on how age modifies the control and regulation of these types of movements. After review of these data on discrete tasks, we focus our discussion on continuous tasks, examining intersegmental dynamics. As coordinated movements involve multiple joints, we consider how the control changes in older adults. In the ensuing section, we shift our focus to whether the elderly can improve their skill level via practice. Two categories of studies have addressed skill acquisition in the elderly. One group of studies shows that the elderly benefit from practice in terms of speed, accuracy, and resource allocation, although the benefits are much smaller than those of the young. A second cluster of studies examined whether the elderly alter the kinematic substructures of movement with practice. These studies suggest that the substructures underlying movement do not change much with practice. Finally, we review a series of studies revealing that older adults become more reliant on vision. While research on the elderly has been successful in documenting performance differences, it has not provided much insight into why these performance decrements occur.

Movement Characteristics

Researchers characterize performance by measuring duration, kinematic, and kinetic variables. In this section we address changes that occur in the perfor-mance of movement of elderly adults.

Movement Duration

Movement time is increased in older adults for a variety of tasks, includ-ing point-to-point movements (Amrhein, Goggin, & Stelmach, 1991; Birren, 1974; Cerella, 1985; Cooke, Brown, & Cunningham, 1989; Ketcham et al., 2002; Goggin & Meeuswen, 1992; Salthouse, 1985), reaching and grasp-ing movements (Carnahan, Vandervoort, & Swanson, 1998; Bennett & Castiello, 1994), handwriting (Amrhein & Theios, 1993; Dixon, Kurzman, & Friesen, 1993; Contreras-Vidal, Teulings, & Stelmach, 1998), and continuous movements (Greene & Williams, 1996; Pohl, Winstein, & Fisher, 1996; Wishart et al., 2000; Ketcham, Dounskaia, & Stelmach, 2001a). Movement durations are on the order of 30% to 60% (50-90 ms) longer in older adults compared to young adults in tasks ranging from simple to complex (Welford, 1977), while in extreme cases, slowing has been reported to be as great as 69% in a point-to-point movement (Stelmach & Goggin, 1988). Movement time is an important measure that captures the general state of the motor system. However, it provides little information about the neural mechanism responsible for the slowing.

Kinematic Profiles

Modern data acquisition techniques make it possible to record and reconstruct movements in real time, permitting investigators to decompose a movement trajectory to obtain information on how a movement is controlled and regulated. It is assumed that the kinematic features are good candidates for what the central nervous system controls as they determine why movement is either faster or slower. Trajectory profiles are processed to create velocity and acceleration profiles, which are further segmented into acceleration and deceleration phases and also parsed into movement substructures (figures 4.2 and 4.3).

Experiments that have employed these kinematic analyses have provided insights into how the movements produced by the elderly differ from those of the young. It has been shown that the velocity profiles of young adults are typically bell-shaped, with the acceleration phase equaling the deceleration phase (see figure 4.2). Studies that have examined trajectory profiles of young and older adults have shown that for the elderly the trajectories are asymmetrical. The deceleration phase is considerably longer than the acceleration phase (Ketcham et al., 2002; Bennett & Castiello, 1994; Brown, 1996; Cooke, Brown, & Cunningham, 1989; Darling, Cooke, & Brown, 1989; Goggin & Stelmach, 1990; Marteniuk et al., 1987; Pratt, Chasteen, & Abrams, 1994; Slavin, Phillips, & Bradshaw, 1996). Brown (1996) conducted an analysis of the ratio of acceleration phase to deceleration phase (1 constitutes a symmetric bell-shaped velocity profile). The range of ratios in older adults was 0.61 to 0.78 whereas young adults had ratios ranging from 0.78 to 0.96. Cooke and colleagues (1989) found acceleration-to-deceleration ratio to be lower in older adults as well and suggest that older adults do not have deficits in the initiation of movements but more in the control of the deceleration phase.

It has been suggested that the deceleration phase contains the portion of movement that is under corrective control since there is sufficient time for sensory feedback to be processed and implemented into the control of the terminal phase of the movement. From a measurement and evaluation standpoint, the deceleration phase in older adults is on the order of 20% to 40% longer than that of young adults (Brown, 1996; Cooke, Brown, & Cunningham, 1989; Pratt, Chasteen, & Abrams, 1994; Bennett & Castiello, 1994; Morgan et al., 1994).

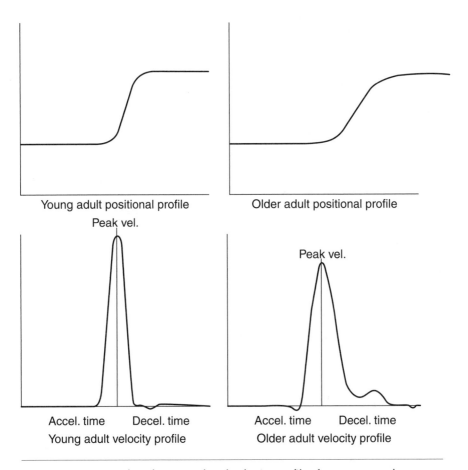

FIGURE 4.2   Examples of positional and velocity profiles for a young and an elderly individual.

Adapted from C.J. Ketcham, R.D. Seidler, A.W.A. Van Gemmert, and G.E. Stelmach, 2002, "Age related kinematic differences as influenced by task difficulty, target-size, and movement amplitude," *Journal of Gerontology: Psychological Sciences and Social Sciences, 57B,* P54-P64. Copyright © The Gerontogical Society of America. Adapted by permission of the publisher.

In addition to longer deceleration phases, the elderly produce movements with 30% to 70% lower peak velocity compared to young adults (Ketcham et al., 2002; Bellgrove et al., 1998; Cooke, Brown, & Cunningham, 1989; Goggin & Meeuswen, 1992; Pratt, Chasteen, & Abrams, 1994; see figure 4.2). Furthermore, when movement distance increases, older adults do not increase the velocity of their movements to the same degree as young adults (Ketcham et al., 2002; Gutman et al., 1993). For example, Ketcham and colleagues (2002) found that the peak velocity of a shorter-distance movement was 15.9 cm/s in elderly and 29 cm/s in young subjects. When movement distance was increased from 9.6 cm to 19.2 cm, the peak velocity of elderly subjects was 27.6 cm/s whereas for young subjects it was 48 cm/s.

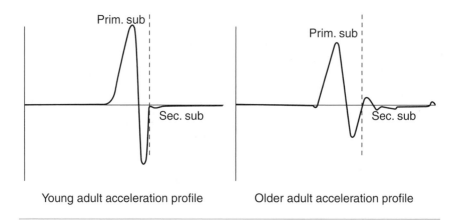

Young adult acceleration profile   Older adult acceleration profile

FIGURE 4.3   Example of acceleration profile and substructure parsing for a young and an elderly individual.

Adapted from C.J. Ketcham, R.D. Seidler, A.W.A. Van Gemmert, and G.E. Stelmach, 2002, "Age related kinematic differences as influenced by task difficulty, target-size, and movement amplitude," *Journal of Gerontology: Psychological Sciences and Social Sciences, 57B,* P54-P64. Copyright © The Gerontogical Society of America. Adapted by permission of the publisher.

For more in-depth analyses of the kinematics of movement, acceleration profiles have also been partitioned into primary and secondary submovements. The primary submovement is the portion of the movement that occurs before the first zero crossing of acceleration, and the secondary submovement is the portion following. The movement optimization model (Meyer et al., 1988) maintains that the primary submovement, in which the limb is propelled to the target during the acceleration phase, represents the portion of the movement under preplanned control, whereas the secondary submovement represents the feedback-controlled portion of the movement (see figure 4.3). The closer to the target the primary submovement ends, the more efficient the motor system is thought to be at that moment (Meyer et al., 1988).

Studies using this analysis have demonstrated that older adults cover 10% to 70% less distance with their primary submovement compared to young adults, depending on the task (Hsu et al., 1997; Ketcham et al., 2002; Pratt, Chasteen, & Abrams, 1994; Walker, Philbin, & Fisk, 1997; Romero et al., 2003; Seidler-Dobrin & Stelmach, 1998). Pratt and colleagues (1994) found that elderly subjects covered 50% of the distance to the target with the primary submovement; in contrast, younger subjects traveled 70% of the total distance (see figure 4.4). Since their primary submovement ends farther from the movement endpoint, the elderly need to make one or more corrective adjustments with the secondary submovement to arrive at the target (Goggin & Meeuwsen, 1992; Hsu et al., 1997; Ketcham et al., 2002; Pohl, Winstein, & Fisher, 1996; Pratt, Chasteen, & Abrams, 1994; Seidler-Dobrin & Stelmach, 1998; Walker, Philbin, & Fisk, 1997). Submovement analyses have been used to determine whether or not the underlying structure of the movement differs between younger and older subjects.

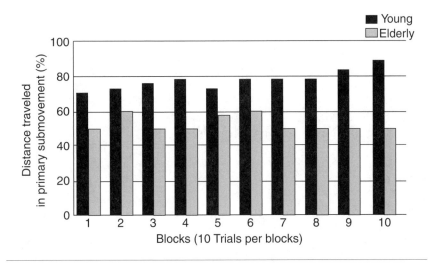

FIGURE 4.4   Proportion of distance traveled in the primary submovement for elderly and young adults in a point-to-point aiming task over 10 blocks of 10 trials.
Adapted from J. Pratt, A.L. Chasteen, and R.A. Abrams, 1994, "Rapid aimed limb movements: Age differences and practice effects in component submovements," *Psychology and Aging, 9* (2), 325-334.

Force Control and Regulation

Since smooth and accurate movements require efficient modulation of force outputs, force control is considered a fundamental component of movement production. Elderly compared to young adults have a reduced ability to ramp and regulate force, making it difficult to initiate and execute movements quickly and accurately across a variety of tasks (Brown, 1996; Campbell, McComas, & Petito, 1973; Clamann, 1993; Cooke, Brown, & Cunningham, 1989; Darling, Cooke, & Brown, 1989; Davies & White, 1983; Doherty, Vandervoort, & Brown, 1993; Galganski, Fuglevand, & Enoka, 1993; Izquierdo et al., 1999; Larsson & Karlsson, 1978; Milner-Brown, Stein, & Yemm, 1973; Milner et al., 1995; Roos et al., 1999; Singh et al., 1999; Stelmach et al., 1989). Stelmach and colleagues (1989), using an isometric task, demonstrated that the elderly had a reduced range of force production and higher force output variability compared to younger adults. In addition, older adults' rate of force production was substantially slower, as it took 20 ms longer to achieve a force level 45% of their maximum (15 N). Ng and Kent-Braun (1999) documented similar findings with older adults. They reported 60 N lower peak force output in elderly compared to young adults and a 20 ms longer time for force production.

It has also been reported that older adults in isometric force tasks produce multiple bursts of force when they achieve targeted force levels approaching near maximum (Kinoshita & Francis, 1996; Brown, 1996; Galganski, Fuglevand, & Enoka, 1993). This is in contrast to the young, who produce a single burst to the targeted force level. While these irregularities are small and occur over short periods, they do provide some insight into why control and coordination change with advanced age. Changes in force regulation and control have large implications for most functional tasks that require precise force generation and termination—for example, pointing to a target, turning a knob, or picking up a glass. Force irregularities may result from motor unit reorganization and muscle composition changes (Erim et al., 1999; Galganski, Fuglevand, & Enoka, 1993; Hakkinen et al., 1996; Yue et al., 1999; Clamann, 1993; Davies & White, 1983; Milner-Brown, Stein, & Yemm, 1973).

Coordination

Coordination is the ability to temporally regulate several body segments in an efficient manner that results in spatially organized movement. With advanced age, controlling multiple movement components becomes increasingly difficult. Such evidence has been found in a variety of movement types: aiming, reaching and grasping, drawing, handwriting, and bimanual coordination tasks (Bennett & Castiello, 1994; Carnahan, Vandervoort, & Swanson, 1998; Teulings & Stelmach, 1993; Greene & Williams, 1996; Swinnen et al., 1998; Wishart et al., 2000; Ketcham, Dounskaia, & Stelmach, 2004; Seidler, Alberts, & Stelmach, 2002). For example, in reach-to-grasp tasks in which the transport and grasp components require precise regulation, it has been shown that these two components are not tightly coupled. Further, in movements that require precise temporal regulation between the arm and shoulder, the elderly show disproportionate irregularities in angular velocities (Seidler, Alberts, & Stelmach, 2002; Ketcham, Dounskaia, & Stelmach, 2004; Carnahan, Vandervoort, & Swanson, 1998). In this presentation we focus on coordination between linked segments.

Intersegmental Coordination

A desired movement emerges from adjustments of neural signals to the specific mechanical properties of the involved limb segments. Joint interactions are the most pronounced mechanical factor influencing multiarticular movements. The limb segments generate these interactions during motion. Therefore, movements about joints cannot be controlled independently from each other.

Dounskaia and colleagues (Ketcham et al., 2004; Dounskaia, Ketcham, & Stelmach, 2002a, 2002b) have conducted a series of experiments that examined repetitive circle, line, and ellipse drawing in the horizontal plane using shoulder and elbow motion. Since muscle activity must be appropriately timed and of optimal magnitude for such tasks, the investigators postulated that preferred joint combinations would be easier to maintain as movement speed increased. The rationale behind this postulation was that the influence of interactive torques on biomechanically disadvantaged joint combinations places a greater burden on muscle activation. Thus, coordination patterns in the direction of interactive torques should be maintained at higher speeds compared to those that require greater modulation.

This paradigm, in addition to providing insights on coordination in young adults, allows researchers to assess how coordination changes as a result of known neuromuscular decrements in elderly adults. In movements performed at the increasing cycling frequencies, Ketcham and colleagues (2004) have shown that the performance of elderly subjects in these mulitjoint drawing tasks is not as good as that of younger subjects. Figure 4.5 depicts differences between groups for the speed conditions of circle drawing. As speed increases, both groups demonstrate distortions in the circle, tending toward an ellipse tilted right. In addition to some shape distortion, older adults tend to show distortions in the size of the trajectory of the circle. Furthermore, elderly subjects show distortions in the slope of the long diameter as speed increases, whereas young subjects do not.

Qualitative differences are a result of differential changes in control between young and elderly adults. Young and elderly adults exhibit differences in the amplitude of elbow excursion and in the relative phase between shoulder and elbow flexion and extension. Elderly adults tend to keep the amplitude of the elbow constant across increasing speeds, whereas young adults increase its amplitude. Additionally, elderly adults decrease the phase offset between the time of shoulder and elbow flexion and extension, whereas younger adults hold it constant across increasing speeds. The decrease in relative phase suggests that elderly adults are trying to modulate interactive torques, but they provide active control before it is necessary while maintaining elbow amplitude,

FIGURE 4.5   Circle trajectories for an exemplary young and elderly subject across increasing movement speeds.

Reprinted, by permission, from C.J. Ketcham, N.V. Dounskaia, and G.E. Stelmach, 2004, "Age-related differences in the control of multijoint movements," *Motor Control*, 8(4): 427.

which leads to distortions. In contrast, young adults do not change the relative phase but rather increase elbow amplitude and apply active control too late to compensate for interactive torques leading to trajectory distortions. These results represent some of the problems the elderly experience when required to control and regulate multiple joints.

Skill Learning

The ability of older adults to acquire new skills or relearn previously learned skills is an important area of motor behavior research. Overall, it has been shown that older adults are able to relearn old skills and acquire novel skills, but they do so at a much slower rate than young adults (Spirduso, 1995; Murrell, 1970; Salthouse, 1984). Skilled abilities such as coordination, balance, associative learning, and handwriting all improve with extended practice, although not often to the level of young adults (Czaja & Sharit, 1998; Harrington & Haaland, 1992; Lazarus & Haynes, 1997; Strayer & Kramer, 1994; Judge et al., 1995; Dixon, Kurzman, & Friesen, 1993; Woollacott, 1993). Older adults require more time to practice a skill before improvements are shown. For example, Judge et al. (1995) found that older adults were able to improve their balance, measured by postural sway range and recovery, when vision and proprioception were distorted, but required substantial practice. Similar improvements were found for recovery of balance with a postural perturbation task after substantial practice (Woollacott, 1993).

Improvements with practice have also been documented with speed of handwriting. Dixon, Kurzman, and Friesen (1993) found that older adults were almost two times slower than young adults prior to practice, but that after extensive practice they were just as fast as the young adults. In another example of skill acquisition on a novel task, Etnier and Landers (1998) demonstrated on a mirror star tracing task that older adults learned at a much slower rate than young adults as measured by time on target. Furthermore, older adults who had more practice also had better retention. Overall, these data suggest that older subjects are able to learn and improve new skills.

Aiming tasks have been used to examine whether the elderly are able to modify the structure of the movements they produce. For example, Seidler-Dobrin and Stelmach (1998) and Pratt, Chasteen, and Abrams (1994) had participants perform an aiming task with accuracy requirements. Both of these studies showed that with extensive practice, overall performance, measured by movement time, improved. However, older adults did not modify the sub-structures of their movements with extensive practice, whereas young participants did. Primarily with practice, the young subjects were able to lengthen the distance traveled with the primary submovement so that it ended closer to the target. Due to the limited modifications in movement substructure, it appears that the elderly are able to make limited modifications in the factors that underlie movement (Cerella, Poon, & Williams, 1980; Seidler-Dobrin, He, & Stelmach, 1998; Brown, 1996; Darling, Cooke, & Brown, 1989; Murrell, 1970). Furthermore, the findings suggest that improvements may occur in different ways than in young adults, which has implications for motor performance in older adults.

There is some evidence that having older adults use cognitive strategies, specifically verbalizing a sequence of content-specific cues, improves the speed of learning (Greenwood, Meeuwsen, & French, 1993; Proteau, Charest, & Chaput, 1994). These strategies help speed up the learning process, make learning more enjoyable, and therefore result in lower attrition rates of older adults in tasks that are novel.

Other data show more specific improvements in cognitive/motor function with specialized training. Kramer, Hahn, and Gopher (1999) have shown very specific benefits from training over several sessions on a dual-task paradigm. They found that older adults, compared to young adults, have large time costs, measured by reaction time, when performing tasks that require switching of attention from one task to another. However, with a modest amount of practice, older adults were able to reduce the costs of switching between tasks. These improvements were maintained over a two-month period. These data show that highly specialized training can improve a very specific kind of performance.

Visual Monitoring

It has been thought that reductions in central planning, proprioception, and force production and regulation, as well as irregular muscle patterns in the elderly, reduce the capability to control volitional actions, thereby creating a need to use vision to guide and regulate movements, particularly during the terminal phase (Seidler-Dobrin & Stelmach, 1998; Slavin, Phillips, & Bradshaw, 1996; Yan, Thomas, & Stelmach, 1998; Larish & Stelmach, 1982; Chaput & Proteau, 1996; Gottlob & Madden, 1999). Thus, visual monitoring of an ongoing movement is thought to compensate for sensorimotor information loss during execution. The consequence of the increased reliance on vision is that movements are typically performed more slowly, have more variability, and exhibit prolonged deceleration phases. Haaland and colleagues (1993) found that when visual information regarding arm position during movements was removed, the performance of older adult subjects on an aiming task was impaired compared to that of young subjects. These results showed that older adults increased movement duration and endpoint errors.

Chaput and Proteau (1996) reported similar findings in a point-to-point movement task. Older adults performed significantly differently when visual feedback was not available compared to when it was; movement times were 100 ms longer, and movement errors were on average 9.3 mm larger. These results have also been documented in other point-to-point movement tasks

(Seidler-Dobrin & Stelmach, 1998; Slavin, Phillips, & Bradshaw, 1996; Yan, Thomas, & Stelmach, 1998; Larish & Stelmach, 1982), in a continuous tapping task (Pohl, Winstein, & Fisher, 1996), and in postural stability tasks (Woollacott, 1993; Whipple et al., 1993; Peterka & Black, 1990). Since visual processing requires continuous online adjustment of a movement, it is often assumed that this is the reason older adults produce slower, more variable movements (Seidler-Dobrin & Stelmach, 1998; Goggin & Stelmach, 1990; Walker, Philbin, & Fisk, 1997).

Support for increased reliance on vision to guide a movement comes from studies that have deprived subjects of complete or partial vision of the limb during movement to a visible target. The kinematics of the movement are then examined to determine whether they are altered in the absence of vision (Seidler-Dobrin & Stelmach, 1998; Darling, Cooke, & Brown, 1989; Warabi, Noda, & Kato, 1986). Seidler-Dobrin and Stelmach (1998) investigated reliance on visual feedback in older adults after substantial amounts of practice and found that practice did not reduce a subject's dependence on vision. Subjects were instructed to move as fast and accurately as possible for 180 trials on a point-to-point movement task. Blocks of trials consisted of a pretest block with disrupted vision, an extensive practice block with vision, and finally a posttest block to see how practice influenced the kinematic structure of the movement when vision was again disrupted. The results documented that both young and older adults relied on visual feedback in the early practice trials. After considerable practice, the young subjects were able to lengthen the proportion of the movement that initially propelled the limb to the target, thus showing their reduced reliance on vision. In contrast, older adults did not demonstrate such a benefit from practice, as they were not able to extend the length of the initial ballistic phase of the movement; consequently they remained highly dependent on vision to guide a greater proportion of the movement.

The studies reviewed provide evidence that older adults are much more dependent on vision to guide and control a movement. The reasons for increased dependence on visual monitoring are not well understood; however, several investigators have suggested that they likely involve limitations in central planning, force production and regulation, proprioception, sensorimotor integration capabilities, or some combination of these (Pohl, Winstein, & Fisher, 1996; Seidler & Stelmach, 1995a, 1995b; Stelmach & Sirica, 1986).

Concluding Remarks

When the control and regulation of movements are examined, research comparing young and elderly participants has consistently documented the decrements in the elderly. These findings have revealed many important performance differences between young and elderly persons. Since these differences have not been very informative with regard to the reasons why such changes occur, the focus of the research agenda needs to change in order to elucidate some of the reasons why the performance of the elderly declines.

In this chapter we suggest that research on the elderly needs to shift its focus to address more of the underlying mechanisms that contribute to the movement decrements observed. This can be done only through comprehensive examination of the underlying mechanisms that determine how movements are controlled and regulated. Some information can be gained through analyzing performance decrements at different levels of control, eliminating or controlling for behavioral strategy differences between groups, determining if the decrements observed are primary or are caused by secondary influences, and establishing neural correlates of observed decrements. Considering these issues will require more complete analyses that scrutinize control at several levels so that researchers can begin to better understand how aging processes alter the control of movement. The advantage of this focus is that it drives research to more closely examine the mechanisms underlying motor control so that performance differences can be more fully understood. However, achieving this goal will not come easy, as the measurement and evaluation issues are considerable.

Chapter 5

Environment, Culture, and Physical Activity of Older Persons

Uriel Cohen, DArch
Ruth Cohen, PhD

This paper examines activities—with particular focus on active living—of older persons, as viewed from a cultural perspective. This perspective has a special significance for elders who suffer from physical and cognitive frailty.

The main goal of this paper is to add to and expand the professional universal concept of physical activities. While the most frequently mentioned activities—walking, biking, and structured exercise—are very effective contributors to physical and mental health, a broader range of other activities can add to, or even replace, those "universal" activities. Our premise is that local, culture-based resources and lifestyle can be the basis for productive and meaningful physical activities, programs, and environments. The paper illustrates our main points through the use of a case study of one cultural community in Bethel, a small rural village in western Alaska.

Culture: A Working Definition

Cultural heritage is defined here as lifestyle, activities, rituals, traditions, values, world view, language, material culture, and other assets and habits shared by a group. This broader, more inclusive definition identifies "cultural groups" not only by ethnic, racial, or religious characteristics; lifelong, shared experiences

Material for this paper was generated in a project sponsored by the Alaska Commission on Aging. The authors would like to thank the people of Bethel, Nome, and Kodiak who have contributed their time and wisdom to this study.

can be a common heritage of cultural groups such as military veterans, retired nuns, or elderly farmers in Minnesota.

Active Living: A Working Definition

Active living (AL) is a way of life in which physical activity is valued and integrated into daily routines and life events. Active living augments or replaces traditional prescriptive exercise programs or structured physical activities in institutional settings, which often turn people off. Traditional approaches to fitness prescribe exercise programs to be conducted several times a week; the level of vigor has to keep one's heart rate up in a target zone.

Active living is a more inclusive concept than physical fitness or exercise. Active living stresses the importance of doing activities that feel good, that are moderate, and that are being perceived as fun. Physical activities are an integral part of daily living—for example, gardening, taking the dog for a walk, or walking the grandkids to school. Thus the contexts for AL might expand to include recreational/leisure activities, social activities, instrumental domestic activities, instrumental vocational activities, and the like.

Benefits of Active Living: Contributions to Physical and Mental Health

Many health benefits and improvements are attributed to AL, such as

- increased energy and physiological well-being,
- maintenance or reduction of body weight,
- increased muscular strength and endurance,
- reduction of back pain, and
- reduced blood pressure and lower risk of coronary heart disease and diabetes.

Many significant social and psychological benefits are also associated with AL, including

- reduction in stress, anxiety, and mild or moderate depression;
- maintenance of functional independence;
- reduced isolation and new opportunities for social interaction and engagement; and
- increased self-esteem and confidence.

The Problem: Barriers to Active Living

Aging presents universal age-related losses that contribute to limitations in physical activity. Frail elders in institutional settings are further limited by organization-imposed constraints. Finally, elders in rural communities, as in the Alaskan village of the case study in this paper, also face local and seasonal barriers to an active lifestyle.

1. The prevailing Western orientation of caregiving and services is grounded in the "medical model," which supports sedentary activities and typically overlooks opportunities for motor activity and outdoor engagement. For example, a typical adult day care center is housed in the proverbial "church basement"—one large space with a group as large as 50 or 60 clients. Typically people sit in a horseshoe pattern and passively watch an "activity" director trying to cheer the sleepy crowd. Economic reasons and other factors further contribute to a limited inventory of activities, superficial engagement, and little variety. As in bad child care, TV watching is a prime-time "activity."

2. Care providers have a limited understanding of cultural resources as potential catalysts for meaningful physical activities. This leaves the universal set of Western-oriented activities (arts and crafts, birthdays, bingo) as the common menu of the daily schedule.

3. Increasing physical and cognitive frailty among the elders creates additional obstacles for productive engagement. Elders' shrinking range of abilities and limited cognitive capacity require greater creativity and commitment on the part of caregivers for generating and maintaining productive solutions. However, creativity and commitment are typically in short supply. Increasing frailty seems to give the overburdened care providers a license for easy "solutions"—for example, letting residents watch TV or take long naps on the couch.

4. In the particular case study discussed in this paper, extreme weather conditions in winter and dark, short winter days put another damper on potential outdoor and motor activities of the elders.

The Premise: Cultural Heritage As a Catalyst for Active Living

The conceptual approach employed here is based on the premise that local, culture-based resources and lifestyle can be the foundation for productive and meaningful activities, programs, and related environments (Cohen & Moore, 1999; Day & Cohen, 2000). Informal studies and anecdotal information indicate that this approach allows for more effective and more meaningful integration of physical activity into programs and the daily life of the community's elders. When these types of activities are compared with large-group, structured, standardized activities common in adult day care centers or in nursing homes, qualitative differences are apparent.

The culture-based approach for programming and design has a direct bearing on the elders' quality of life in several domains:

- Maintenance of functional independence
- Increasing personal autonomy
- Meaningful social contact
- Continuity of self (Cohen & Weisman, 1991)

The Context of the Case Study: History, Culture, Demographics, and Economy

The geographical setting of the following case study is the village of Bethel, Alaska, and several smaller outlying villages. The primary user group addressed was the community elders at their homes and in independent living housing, an adult day care center, a senior center, and assisted living facilities.

Bethel was first established by Yupik Eskimos, who called the village "Mumtrekhlogamute," meaning "Smokehouse People," the name referring to the nearby fish smokehouses. Today 80% of the population are Alaska Natives. A federally recognized tribe is located in the community. The traditional Yupik Eskimo practices and language remain predominant in the area. Subsistence activities and commercial fishing are major contributors to residents' livelihoods. Subsistence activities contribute substantially to villagers' diets, particularly with regard to salmon, freshwater fish, game birds, and berries.

Unique Cultural Resources or Assets of the Community

Culture-based and local activities in Bethel include instrumental activities related to subsistence lifestyle; food gathering, hunting, fishing, and food processing/preserving; activities of daily living; activities associated with rituals, traditions, and special events; and activities related to performing arts, crafts, and expressive activities.

Illustrative Case Study

The case study was a part of a technical assistance project sponsored by the Alaska Commission on Aging. The project examined 12 adult day care centers across Alaska to evaluate their responsiveness to dementia-specific issues, leading to design recommendations for improvements. Examining the subject from a cultural perspective was part of our assessment methodology, since we felt that potential interventions might be shaped by local cultural factors.

Figures 5.1 through 5.4 depict one example of activity based on simple but distinctively local cultural phenomena. The local tradition of subsistence living,

FIGURE 5.1   A frail Yupic elder gathering greens in the tundra near Bethel, Alaska.

FIGURE 5.2   Adult day care clients hard at work in the fields.

FIGURE 5.3   Back at the senior center, the daily harvest is being sorted and processed by the elders.

FIGURE 5.4   Dora M., 94, obviously proud of her harvest of wild berries following a full afternoon's outing.

Courtesy of Nome Alaska Adult Day Care.

which centers on gathering and processing food, was a catalyst for several activities. In this particular example, a group of adult day care clients went for an outing to nearby fields. The elders spent over 2 hr walking in the field gathering greens (figures 5.1 and 5.2). Upon their return to the senior center, the elders continued the activity in the dining room, sorting and processing their harvest (figures 5.3 and 5.4).

This illustration is one of numerous examples of culture-based activity specific to one cultural group. The case study described here indicated that motivational aspects of activities are among the key predictors of success in engaging the elderly in AL or the failure to do so. Findings indicate that elders will respond more positively to activities that are not perceived by patrons to be institutional or an "exercise program." One positive approach to AL is to create culture-based programs and environments that accommodate familiar, ordinary, and enjoyable activities integrated into everyday life. Cultural assets can provide the contents and foundation for these activities.

Conclusions

The discussion and case study explore alternative approaches for AL: culture-based physical activity or "stealth exercise." Our premise is that *meaningful* activities reflect an interaction between a person with a prior history of experience and a task environment in which that history is relevant.

Some of the culture-relevant dimensions may contribute to improved performance and more positive experiences. For example,

- the cultural context may offer *activity-motivating* factors that serve as a trigger or a catalyst to action;
- the cultural context may offer *performance-enhancing* factors contributing to involvement, engagement, and sustained interest; and
- culturally relevant activities may have an *enhanced affect*—affirmation, satisfaction, personal meaning, and improved self-image

The challenge is to develop appropriate indicators, measures, and methodologies through which the unique aspects of the culture-based approach to activities can be evaluated and assessed.

Part II

Measurement Challenges in Aging Research

Chapter 6

Physical Activity, Aging, and Quality of Life

Implications for Measurement

Edward McAuley
Steriani Elavsky

Older adults are the most rapidly growing segment of the U.S. population, with approximately 35 million adults aged 65 years and older, a number that is projected to double in the next several decades (U.S. Census Bureau, 2000). With advancing age there is an increased susceptibility to various chronic conditions, functional disability, and comorbidity, often resulting in compromised physical, emotional, and psychological well-being and reduced quality of life (QOL). Physical activity interventions represent an effective behavioral strategy not only for attenuating functional decline and reducing risk of disability (Miller et al., 2000; Singh, 2002; Keysor, 2003), but also for enhancing psychological well-being and QOL in older adults (Rejeski, Brawley, & Shumaker, 1996; Stewart et al., 1997; McAuley et al., 2000).

The determination of whether physical activity effectively enhances QOL is predicated on our being able to accurately operationalize and reliably measure this ubiquitous construct. In this chapter, we begin by discussing how QOL has been conceptualized and defined and then present an overview of the typical types of measures that have assessed QOL in general. We follow this with a review of the physical activity and QOL literature. Given space constraints we summarize several contemporary reviews that examine this relationship with

Preparation of this chapter was facilitated by support from the National Institute on Aging (Grants AG 12113 and AG 20118).

respect to aging, cancer, and dose-response issues (Courneya & Friedenreich, 1999; Rejeski & Mihalko, 2001; Schechtman & Ory, 2001; Spirduso & Cronin, 2001). We devote the remainder of the chapter to issues that we believe can be tackled by measurement specialists and that have important implications for subsequent study of the physical activity and QOL relationship in older adults.

Conceptualizing and Defining Quality of Life

Rejeski and Mihalko (2001) have correctly identified a lack of precision in defining QOL as a major hurdle in making consensus statements relative to the relationship between physical activity and QOL. In defining QOL, it is important to realize that a true conceptualization of this construct must take into consideration both the subjective nature of the evaluation of QOL and the process in which one compares one's current life with some personally identified criteria (Trine, 1999). In so doing, it becomes apparent that many conceptualizations and definitions of QOL fail in this respect.

For many years, QOL was assessed as a function of morbidity and mortality indices. Thus, the absence of disease or life expectancy predictions were considered as a marker of QOL. These indices were further augmented in the medical literature by the assessment of functional status to determine how illness and prescribed treatment of disease influenced overall health status or health-related quality of life (HRQL; Rejeski, Brawley, & Shumaker, 1996). The assessment of functional status as an indicator of QOL has operated under the premise that having compromised physical function, that is, being unable to perform certain activities of daily living, is associated with degradations in QOL (Thomas, 2001). Such a limited perspective ignores the adaptability of humans, as individuals with considerable physical limitations also report relatively high levels of life satisfaction (Thomas, 2001).

Many researchers have taken the position that QOL is a multidimensional or umbrella construct under which such dimensions as physical, social, psychological, and spiritual well-being reside. Of particular relevance to the physical activity and aging literature is the conceptualization of QOL offered by Stewart and King (1991). This framework views QOL from the perspective of two broad categories, function and well-being, which are subserved by multiple, more specific, QOL outcomes. Two issues relative to such a framework are worth considering. One concerns the extent to which such outcomes are important to or valued by the participant. The second issue concerns the extent to which some of these elements that are classed as QOL outcomes are indeed representative of QOL. For example, it is well established that physical activity is associated with decreases in depression and anxiety and enhanced self-efficacy and self-esteem (see McAuley & Katula, 1998 for a review). However, as already noted, as well as reflecting a subjective representation, QOL also involves comparative assessment (Trine, 1999). Clearly, the aforementioned variables and many other constructs conceptualized as QOL do not meet both of these criteria.

Adopting Diener's (1984; Pavot & Diener, 1993) position that QOL is a "cognitive judgment of satisfaction with one's life," Rejeski and Mihalko (2001) have made perhaps the strongest statements about the need to consider QOL, in relation to physical activity and aging, at the level of the psychological construct. Such an approach allows for comparative judgments, places importance on cognitive assessments, and has implications for theory testing and development (Rejeski & Mihalko, 2001). Thus, assessment of QOL at the global level, and in particular as satisfaction with life, would relegate many commonly assessed constructs (e.g., anxiety, depression, esteem, pain, physical function) to the level of intermediate QOL outcomes that underlie global constructs such as satisfaction with life.

Measuring Quality of Life in Physical Activity Research

A recent bibliographic study by Garratt et al. (2002) identified 3,912 studies in which 1,275 patient-assessed measures of QOL were developed or evaluated. In an earlier paper, Stewart and King (1991) classified QOL measures that have specifically been used in the physical activity domain along several categories. *Aggregate measures* rely upon combining multiple categories of QOL (e.g., physical symptoms, physical function, depression, control) to arrive at a single index of QOL. Clearly, this type of assessment is likely to underestimate the effects of physical activity on QOL, as such effects may be attenuated when combined with outcomes unlikely to be influenced by physical activity. Stewart and King also note that the tendency of some researchers to employ *single-item* measures of QOL is problematic due to their lack of reliability, validity, and sensitivity. The tendency to use such measures appears to have been reduced in recent years, but QOL measures may still take this form in larger epidemiological studies.

When QOL is measured as *perceived change* in health outcomes associated with an intervention, there is, of course, the likelihood that the demand characteristics of the questions will lead to a response bias on the part of the respondent. When this measure is used in conjunction with objective measures of change, this bias can be ameliorated somewhat (Stewart & King, 1991). Citing the Medical Outcomes Study (MOS) 36-item Short Form Health Survey (SF-36; Ware & Sherbourne, 1992), Stewart and King make a strong case for the use of *multidimensional measures* of QOL that employ assessments of the primary domains of function including physical, mental, and social function. The advantages of such an approach lie in being able to document change in QOL at varying levels that may be more susceptible to change through physical activity.

It is important to extend the categories laid out by Stewart and King (1991) to draw attention to measures that might be termed *nonmeasures of QOL* and that are represented by constructs such as mood, anxiety, and depression. These constructs are often assessed, and QOL is inferred from their scores. Although these constructs are undoubtedly related in some way to more global assessments of QOL, they are theoretically and conceptually distinct from QOL.

Finally, we would argue, in much the same way that Rejeski and Mihalko (2001) have done, that measures designed to allow the respondent to decide which components of one's life are important in making judgments relative to QOL best reflect the self-referenced subjective nature of global QOL and are better suited for theory testing and development. The Satisfaction with Life Scale (Diener et al., 1985) is a good example of such a measure and targets life as a whole rather than identifying particularized facets or domains.

Can Physical Activity Improve Quality of Life in Older Adults?

We now consider the extent to which physical activity is related to QOL in older adults. In general, this is an area that has witnessed a considerable increase in descriptive, prospective, and clinical trials resulting in several recent and thorough reviews (e.g., Courneya & Friedenreich, 1999; Rejeski & Mihalko, 2001; Schechtman & Ory, 2001; Spirduso & Cronin, 2001). Given space constraints, we present a general overview of what these reviews have concluded. The four reviews that best capture the physical activity and QOL relationship take quite different approaches to the topic and thereby provide unique information on this relationship. For example, Rejeski and Mihalko (2001) organize their review relative to studies examining the relationship in the context of multidimensional or umbrella definitions and measures, and offer their perspective on possible moderator and mediator variables that might explain the physical activity and QOL relationship in older adults. Spirduso and Cronin (2001) focus on the issue of dose response, as do Schechtman and Ory (2001), although the latter do so by employing preplanned meta-analyses of the FICSIT (Frailty and Injuries: Cooperative Studies of Intervention Techniques) randomized clinical trials. Finally, the increasing interest in employing physical activity as a palliative intervention in clinical disease populations led us to include a review of the cancer, physical activity, and QOL literature by Courneya and Friedenreich (1999).

In an earlier review of the physical activity and QOL literature, Rejeski, Brawley, and Shumaker (1996) focused entirely on those studies that had embraced QOL from the "umbrella" or multidimensional perspective of assessing multiple outcomes that may represent function and thereby HRQL. Because this approach to QOL presents difficulties with identifying potential mediators and moderators of the relationship, in their updated review they also consider those studies employing measures that consider QOL at the level of the psychological construct, or as satisfaction with life.

Rejeski, Brawley, and Shumaker (1996) initially reviewed 28 studies involving a variety of healthy and diseased samples, reaching the overall conclusion that the relationship between physical activity and QOL in older adults was positive, independent of age, activity, and health status. However, this positive relationship was not apparent in all areas of function. For example, the relationship was nonsignificant or attenuated in older adults who were already functioning at or above the norm. Moreover, there appears to be little relationship between measures of cardiovascular fitness change and improvements in QOL (Rejeski, Brawley, & Shumaker, 1996). Indeed, as Rejeski and Mihalko (2001) cogently point out, performance-based assessments of function are more likely to be related to QOL, as such improvements are much more relevant to older adults than, for example, a 5% increase in aerobic capacity. Moreover, the literature reviewed typically inferred QOL improvements from changes in function and symptoms that were independent of assessments of respondents' perceptions of improvement and satisfaction with function.

In their recent review, Rejeski and Mihalko (2001) considered an additional 18 studies (including 12 randomized controlled trials) that fell under the rubric of QOL as an umbrella construct. They once again confirmed the consistency of the relationship between physical activity and QOL in older adults; and in spite of an array of measures being used to assess multidimensional QOL, this relationship held across subgroups, activity settings, and activity mode.

However, it is with respect to the consideration of QOL at the level of the psychological construct that Rejeski and Mihalko have made some of their strongest statements. In this context, QOL is operationally defined as satisfaction with life. Twelve studies are reviewed, and the findings are generally equivocal, with cross-sectional studies consistently indicating a positive relationship between physical activity and life satisfaction. However, only three of six randomized controlled trials have demonstrated positive physical activity effects on QOL. Rejeski and Mihalko (2001), suggest that a lack of consistency in both exercise prescriptions and measurement of life satisfaction may account for this equivocality (see also McAuley & Rudolph, 1995; Rejeski, Brawley, & Shumaker, 1996).

Rejeski and Mihalko (2001) have stated that examining QOL at the level of the psychological construct, and particularly as satisfaction with life, provides opportunities for testing and developing theory. One of the ways in which this can be done is to determine which factors may moderate and/or mediate the relationship between physical activity and QOL. Citing the consistently reported associations between physical activity and self-related constructs such as self-efficacy and positive feeling states (e.g., McAuley & Katula, 1998) and self-esteem (e.g., McAuley & Rudolph, 1995), Rejeski and Mihalko propose that such constructs as enjoyment, self-efficacy, and other self-schema may be important mediators between physical activity and QOL (satisfaction with life) in older adults. Self-efficacy is the active ingredient in Bandura's (1986) social cognitive theory and has been widely applied in the physical activity and psychological outcomes literature.

Certainly, there is good experimental evidence, albeit with young adults, to suggest that manipulating self-efficacy in physical activity contexts differentially influences feeling-state responses such as psychological distress, positive well-being, and fatigue (e.g., McAuley, Talbot, & Martinez, 1999; Jerome et al., 2002) and anxiety (Marquez et al., 2002). In addition to potential mediating variables, Rejeski and Mihalko also suggest that the relative value that older adults place on functional abilities and on physical activity itself may well moderate any effects of physical activity on QOL. This becomes a particularly important consideration when one is reviewing the greater QOL literature, in which, all too often, no consideration is given to the value or importance placed on outcomes identified as QOL related. We would argue that many of these QOL outcomes are intermediate outcomes that may mediate activity effects on global QOL and may also be moderated by personal value systems. Further examination of the mediators and moderators of the physical activity and QOL relationship is warranted.

Can Physical Activity Improve Quality of Life in Cancer Patients?

We turn now to a rapidly expanding literature in the area of physical activity and QOL across the cancer experience. The importance of studying physical activity and QOL relationships in this population is directly related to the fact that increased incidence rates and improving survival rates have resulted in approximately 9.8 million Americans living with cancer, with about 1,372,910 new cancer cases expected to be diagnosed in 2005 (American Cancer Society, 2005). Living with cancer means prolonged medical treatments such as radiation, chemotherapy, or surgery; and such treatments serve to compromise psychological, emotional, and functional well-being. In short, these treatments can have a profound effect on reducing QOL. Consequently, physical activity may have considerable potential as an intervention to attenuate the effects of cancer treatment from a QOL perspective. Courneya and Friedenreich (1999) provided one of the first comprehensive reviews of physical activity effects on QOL following diagnosis of cancer.

As with much of the physical activity and QOL literature, a host of measures reside under the QOL rubric in cancer-related studies. Overall, Courneya and Friedenreich conclude that following cancer diagnosis, physical activity (either as adjuvant therapy or posttreatment) had a positive influence on an array of QOL outcomes across 16 of 18 intervention studies and five of six descriptive studies. From a functional standpoint, benefits were demonstrated in the context of improved functional capacity, strength, body composition, fatigue, pain, and immune function whereas psychological benefits included improved mood states, anxiety and depression levels, esteem, and satisfaction with life. On the positive side, it appears that these effects are fairly robust, with approximately 90% of the studies showing statistically significant QOL improvements due to physical activity in samples averaging 24 patients per study.

On the negative side, however, there is an overreliance in this literature on the use of physiologic indices as primary outcome measures. When subjective psychological assessments are made, they too often focus on relatively narrow constructs such as anxiety and mood states. As Courneya and Friedenreich (1999) further point out, such approaches fail to consider QOL from a global and comparative perspective. These authors recommend that the study of QOL in the context of exercise and cancer adopt a broader QOL framework and thereby identify which dimensions of QOL are likely to be affected. As noted earlier, the determination of which aspects of function effectively influence QOL is dependent upon the extent to which value or importance is placed on a given domain by the participant. Once again, we would argue that many of the QOL outcomes assessed in the cancer literature could be considered intermediate outcomes that underlie more global QOL assessments representing satisfaction with life.

Is There a Dose-Response Relationship for Physical Activity Effects on Quality of Life?

Knowing the extent to which a particular dosage of physical activity brings about a meaningful change in some health outcome allows clinicians and practitioners to accurately prescribe physical activity regimens and permits governing bodies and policy makers to formulate effective recommendations for activity participation. Two reviews, one by Spirduso and Cronin (2001) and one by Schechtman and Ory (2001), take different approaches to determining whether there is a dose-response relationship between physical activity and QOL. In a traditional narrative review, Spirduso and Cronin followed the Stewart and King (1991) classification model of QOL with function (physical, cognitive, self-maintenance activities, fitness, disease symptoms) and well-being (bodily, emotional, esteem, global) as the umbrella terms under which multiple constructs representing QOL are captured. Schechtman and Ory (2001), from the FICSIT group, explored the issue of dose response by conducting a preplanned meta-analysis of the QOL outcomes from four randomized controlled physical activity interventions.

Overall, the Spirduso and Cronin review suggests little support for a dose-response relationship, and there are several reasons why this may be the case. First, given the disparate nature of QOL assessment in these studies, it is actually unsurprising that no clear pattern of dose response has emerged. Some clinical evidence at the cross-sectional and descriptive level is reported as suggesting a dose-response relationship between quantifiable physical activity outcomes such as strength, muscular power, and fitness and "QOL" outcomes such as chair rising, stair climbing, and walking. However, these latter outcomes represent functional outcomes rather than indicants of some comparative judgment of how satisfied one is with one's life. Moreover, the extent to which individuals consider QOL outcomes of relevance and the value that they place on those outcomes govern the extent to which physical activity is going to influence them.

In the FICSIT group meta-analysis, QOL was assessed by the SF-36 (Ware & Sherbourne, 1992). Initial examinations of whether physical activity was related to QOL revealed that the different types of physical activity interventions improved only the emotional well-being component of the SF-36, with a trend toward improvement in social functioning but no changes evidenced in the general health scale. Interestingly, the interventions had no significant effect on the bodily pain scales. Any such increase would have been interpreted as an adverse consequence of exercise, and therefore Schechtman and Ory (2001) view this latter finding in a positive light. That is, older frail adults (60% with arthritis) did not suffer any *more* pain as a result of physical activity.

In an effort to examine whether a dose-response relationship might exist, the intensity of the interventions was categorized as high, medium, or low based upon the exertion level of the actual activity and kilocalories burned per week as indicated by frequency and duration of activity. Overall, there was no support for a dose-response relationship, and improvements in QOL were unrelated to improvements in physical function as assessed by gait speed. There may be several explanations for such findings. First, gait speed is clearly only one of many aspects of physical function that come into play in older adults' lives. Second, Schechtman and Ory (2001) point out that their findings relative to the lack of a dose response are hampered by the lack of precision in their compliance data and a relatively weak measure of intensity. Further, the authors speculate on whether the SF-36 is sensitive enough to the effects of physical activity in such a sample, suggesting that "more detailed measures of anxiety and depression" may be more sensitive in demonstrating physical activity effects. However, we would contend that measures of anxiety and depression are not measures of QOL and, although related, are conceptually distinct constructs. We believe it is important to recommend that future research place a premium on assessing the level of importance and relevance that is attached to physical activity, as well as the domains assessed as QOL, rather than assuming that the measures used are adequately capturing what it is in their lives that older people value.

Issues to Consider in the Physical Activity and Quality of Life Relationship

From our review thus far, we can conclude that there is a positive relationship between physical activity and QOL that is supported by both intervention and cross-sectional data in older adults (Rejeski & Mihalko, 2001) and cancer patients (Courneya & Friedenreich, 1999). However, inconsistencies abound in the conceptualizations of the QOL construct and the types of measures employed. Finally, there appears to be little support for a dose-response relationship in the quality of life–physical activity literature (Schechtman & Ory, 2001; Spirduso & Cronin, 2001). We now give brief consideration to several issues relative to the relationship of interest that we believe measurement specialists are well placed to address.

Measures

An obvious question may concern whether the field needs better measures of QOL. We remain unconvinced that more measures are necessary. However, we firmly believe that it is important to identify a priori QOL outcomes that are valued and relevant to participants rather than to assess multiple constructs and infer QOL to have been influenced. For example, older adults may not care very much about body composition and aerobic capacity, *but* they do care about maintaining independence, memory, and social ties. It is improvement or maintenance of these aspects of function that have implications for how satisfied one is with one's life. Thus, these functional outcomes can, as we and others (Rejeski & Mihalko, 2001) have suggested, be measured and classified as intermediate QOL outcomes that underlie the more global construct of QOL.

In addition, as we adopt some consistency in the measures used to assess QOL in the physical activity domain, it will be necessary to further determine the psychometric qualities, in particular the construct validity, of these measures. For example, as physical activity interventions for influencing QOL are being more broadly applied, it will be necessary to fully establish the extent to which these measures are invariant across race, ethnicity, subsets of older adults, and so forth. The SF-36 (Ware & Sherbourne, 1992) is an example of a multidimensional QOL measure that has enjoyed careful validation and very broad application.

Degree of Change in Quality of Life

It is also appropriate to ask whether we can expect to see physical activity-related changes in all older adults. For example, older adults who already enjoy a relatively high level of QOL are unlikely to experience substantially enhanced QOL as a function of physical activity participation. A good example of this was reflected in the outcomes of a series of studies by Blumenthal et al. (e.g., Blumenthal & Madden, 1988; Blumenthal et al., 1991) that failed to demonstrate any substantive changes in elements of QOL (in this case, cognitive function) following a physical activity intervention. However, the majority of the participants in these studies were older current or retired faculty members at Duke University who were not only fairly active initially but presumably demonstrated reasonably high levels of functioning.

Alternatively, in samples where health is compromised (e.g., cancer) or function is limited (e.g., the disablement process) and where QOL may be quite dramatically reduced, it is quite possible that physical activity may influence QOL, especially in those realms of QOL that may be important *and* are being affected.

Clinical Importance of Quality of Life Improvements

An important related issue has stemmed from an impressive increase in the number of published studies in the past two years documenting the palliative physical activity effects on QOL across the cancer experience. This increase in research activity has given rise to an important clinical question of how significant any improvements in QOL may be. Although this question has its roots in determining the effects of all treatments on QOL, it is particularly salient for physical activity interventions. Consider, for example, the case of fatigue. This is a common problem associated with cancer and can be debilitating from a QOL perspective. Fatigue can be objectively measured as a function of patients' exercise tolerance, their hemoglobin levels, or, alternatively and very commonly, by self-report. With respect to the latter type of assessment, the fatigue scale of the Functional Assessment of Cancer Therapy (FACT-F; Yellen et al., 1997) is one of the most commonly employed measures of fatigue in this population. However, until recently, it has not been possible to determine to what extent improvements in fatigue scores, as measured by the FACT-F, represent clinically important differences.

To address this issue, Cella and his colleagues (Cella et al., 2002) have employed both anchor- and distribution-based methods to determine the range of score changes in QOL measures (e.g., FACT-F) that correspond to differences or changes in function or clinical course (e.g., hemoglobin levels, performance status, and response to treatment). Their work provides effect sizes for determining how a change score on the FACT-F relates to well-documented, clinically significant change in other objective measures. Given the increase in the application of physical activity as both palliative and adjuvant therapies, it would appear that the establishment of such clinically important differences will take on more significance in future physical interventions. Moreover, such data can serve as important and meaningful information for both patients and their health care providers (Cella et al., 2002).

Mechanisms Underlying the Relationship

Regardless of how QOL is defined and measured, the evidence would generally support the conclusion that physical activity has at least a modest effect on QOL in both healthy and diseased older adults. The most important question may concern the underlying mechanisms that might explain this relationship. As with the larger exercise and mental health literature, numerous physiological, social, biochemical, electrophysiological, and psychological hypotheses have been advanced to explain this effect (Trine, 1999). However, there have been few elaborate examinations of these underlying mechanisms in older adults. Additionally, there is little compelling evidence to suggest that the improvements in QOL are directly brought about by objective improvements in fitness, function, body composition, and so on.

As noted earlier, some evidence exists to support the social cognitive perspective that would propose that effects on life satisfaction, for example, would be influenced by these physical outcomes (as well as other types of outcomes) through the mediation of self-efficacy (e.g., Rejeski & Mihalko, 2001). Self-efficacy, as already mentioned, is the active ingredient in Bandura's (1986) social cognitive theory and has been widely applied in the physical activity and psychological outcomes literature. Efficacy expectations are beliefs in capabilities relative to successfully completing challenging behavioral tasks. Subsequent behavior is reciprocally determined by past behavior, environmental, and interpersonal factors. As QOL is by definition a cognitive judgment, it is very likely that cognitive outcomes (self-efficacy, esteem) associated with improvements in behavioral/physical outcomes brought about by physical activity are implicated in changes in life satisfaction. Further examination of the mediators and moderators of the physical activity and QOL relationship is warranted.

Measuring Change in Quality of Life

Research designs of studies examining physical activity effects on QOL in older adults have become more sophisticated and have more frequently included randomized controlled trials. At the heart of such trials is the interest in assessing change brought about by the physical activity intervention, the potential determinants associated with such change, variations in initial status variables that may moderate change, and the potential for having to deal with hierarchically structured data sets.

To effectively evaluate the extent and determinants of such change, it will be important for measurement specialists to promote, and researchers to adopt, contemporary statistical methods (Masse et al., 2002). For example, latent growth curve modeling (see Duncan et al., 1999 for an overview of growth curve applications) has been successfully incorporated into a number of physical activity interventions involving older adults (e.g., McAuley et al., 2000). Such approaches allow one to consider intervention effects on intraindividual change and interindividual differences in such change. Finite mixture modeling is a further advanced technique that affords researchers the means to examine heterogeneity in growth parameters (Muthén, 2001). Such approaches may prove particularly useful in determining how underlying latent classes of key variables might moderate physical activity effects on QOL. Additionally, researchers have begun to take into account the multiple layers or levels that are inherent in applying interventions to health and physical activity behavior. To effectively analyze such data, multilevel modeling applications (e.g., Bryk & Raudenbush, 1992) are likely to become increasingly attractive in examinations of the physical activity and QOL relationship.

Concluding Remarks

The demographics of our aging society dictate that QOL will continue to be a very important health promotion objective for the nation. The relationship between QOL and physical activity appears to be positive and relatively consistent in older adults, but there is a need to more carefully refine our conceptualization and measurement of the QOL construct. Additionally, there is a great need for determining what the moderators and mediators of this relationship might be. Finally, there is continued need to differentiate between intermediate and global outcomes of QOL and to ascertain that intermediate outcomes are of value and relevance to participants. Unless this is done, it becomes very difficult to determine whether indeed such outcomes truly represent QOL.

Chapter 7

Assessment Issues Related to Physical Activity and Disability

James H. Rimmer, PhD

Innovative strategies for improving health, preventing complications associated with a disability, and adequately preparing individuals with disabilities to understand and monitor their own health have emerged as an important public health priority (Patrick et al., 1994). While people with disabilities compose only 17% of the noninstitutionalized population of the United States, they account for 47% of total medical expenditures (Rice & Trupin, 1996). On average, their medical expenditures are more than four times those of people without disabilities (Rice & Trupin, 1996).

Despite the enormous health benefits that can be derived from regular physical activity (U.S. Department of Health and Human Services, 1999, 2000), people with disabilities remain one of the most physically inactive groups in society (Durstine et al., 2000; Heath & Fentem, 1997; Rimmer & Braddock, 1997; Rimmer, 1999; U.S. Department of Health and Human Services, 2000). The level of physical inactivity observed in this population has been linked to an increase in the severity of disability and erosion of involvement in community activities (Ravesloot et al., 1998). These patterns of low physical activity reported among people with disabilities raise serious concern regarding their health and well-being, particularly as they enter their later years when the effects of the natural aging process are compounded by years of sedentary living and severe deconditioning (Rejeski & Focht, 2002).

As persons with disabilities age, they often experience increasing difficulty performing activities of daily living (ADL) (e.g., dressing, showering) and instrumental activities of daily living (IADL) (e.g., ambulation, doing laundry,

This work was supported, in part, by the National Institute on Disability and Rehabilitation Research, #H133E020715.

grocery shopping) at a much earlier stage compared to the general population of older adults (Rimmer, 2001). The necessity for people with disabilities to maintain their physical ability is, in some respects, even more important than in the general population because of their narrower margin of health. Some experts believe that the high incidence of physical inactivity and poor health practices observed in people with disabilities (Heath & Fentem, 1997; Rimmer, 1999), combined with this natural aging process and the potential loss of function from the disability itself, presents a volatile combination for people living at or close to the "threshold" of physical dependence.

Researchers in recent years have begun to develop interventions to examine the effects of physical activity on reducing secondary conditions and improving overall health among people with disabilities. However, efforts to determine the efficacy of these interventions have been limited by the lack of reliable and valid assessment tools for measuring physical activity in this population. This paper deals with measurement issues related to physical activity and disability, including a brief discussion of the issues surrounding the definition of *disability* and the impact that the *environment* has on participation in physical activity among people with disabilities.

Defining Disability

Defining *disability* is not a simple task. Given that there are various definitions of disability and that many people who have a disability do not consider themselves "disabled" according to the legal definition used in the Americans with Disabilities Act (i.e., two or more functional limitations), the controversy of who *is* and *is not* disabled still exists. Unfortunately, a universally accepted definition has never been decided on in the medical or disability communities, and there are different estimates under varying survey schemes and sampling frames (Zola, 1993). The "approaches" to disability data are best viewed as crude approximations at a point in time, with considerable, but not total, communality. Discrepancies between estimates underscore the fluidity of the disability construct and the vagaries of identification.

Statistical summaries of disability typically employ definitions based on (1) the presence of a physical or health impairment or (2) an assessment of "loss of function." Either concept is common to everyday experience. In practice, however, layers of nuance are revealed. Impairments may not functionally affect an individual's life. Function is not so readily operationalized—which skills or functions are most important? What is the threshold of loss? Are thresholds absolute or relative? To what extent are external accommodations—assistive devices or environmental modifications—incorporated into the definition? This latter question reflects a third perspective best represented by the still-evolving notion of "restricted participation" in life situations in the recently implemented International Classification of Functioning, Disability, and Health (World Health

Organization, 2001). Here, the defining element is the interchange of person-level characteristics within the context of the environment. The difficulty of translating these concepts into easily administered "head counts" is enormous. In fact, under what might be termed "situational disablement" paradigms, the principal measurement unit is not the person, but rather the specific interaction between the person and environment.

Part of the challenge in assessing various constructs in physical activity and fitness among persons with disabilities is identifying how to cluster samples. Individuals with the *same* disability (e.g., multiple sclerosis, cerebral palsy, rheumatoid arthritis) may present vastly different levels of function and impairment, while individuals with varying disabilities (e.g., stroke, head injury, spastic hemiplegia) may present similar levels of function and may be more appropriately grouped by specific functional impairments (e.g., hemiplegia). However, many researchers have difficulty publishing papers or reports that have different disability groups included in the same data set because of the presumption by the reviewers that the sample is too heterogeneous and not generalizable to any specific population.

While sociologists and epidemiologists continue to grapple over what is considered or defined as a *disability*, one thing is clear: The number of children, adults, and seniors who have some type of physical, cognitive, or sensory limitation continues to increase. In 1997 the national prevalence of disability as measured by the National Health Interview Survey (NHIS) was 13.3% of the 266.6 million noninstitutionalized Americans in the population. Under a broader, functional-limits conception in the Bureau of the Census Survey of Income and Program Participation, the 1997 rate was close to 20%, which included 53 million Americans (Fujiura, 2001). As illustrated in figure 7.1, a substantial proportion of the U.S. population self-reports as having a mobility

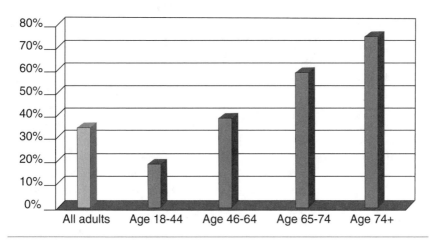

FIGURE 7.1   Percentage of Americans with mobility limitations by age, United States, 1998.

limitation, and millions of individuals (figure 7.2) use an assistive device to help with ambulation (one of the key criteria for having a disability according to the Americans with Disabilities Act).

Establishing viable physical activity interventions to reduce the effects of functional limitations must become a major priority among researchers in exercise science. Measurement of these interventions will require valid and reliable assessment tools that can capture effective changes among persons with varying levels of disability or function (or both) and determine how these changes might relate to improvements in overall health and function (e.g., being able to do more ADL and IADL independently).

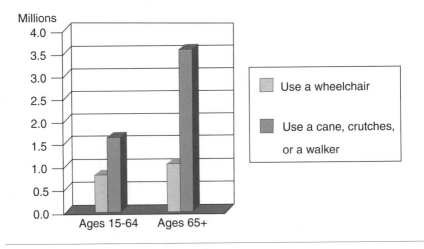

FIGURE 7.2   Number of Americans who use a mobility device by age.

Importance of Physical Activity in Improving Function

The relationship between physical functioning and physical activity is a reciprocal one; physical functioning provides the individual with the capability to engage in physical activities, whereas physical activity helps to maintain and in some cases improve physical functioning. This reciprocal relationship, coupled with the high prevalence of physical inactivity among persons with disabilities, has profound implications for rehabilitation practice, especially in evaluating intermediate- and long-term outcomes of clinical practice. For rehabilitation to play a role, however, in the long-term maintenance and enhancement of physical functioning among people with disabilities, measurement of physical activity must become an essential component of the recovery and maintenance process.

The development of standardized tools for assessing physical activity levels of people with disabilities can serve the purpose of monitoring physical activity patterns at both the individual (clinical application) and population (public health surveillance) levels. Presently, there are no valid, reliable, and user-friendly assessment tools that can capture levels of physical activity after a patient is discharged from a hospital or long-term care facility. While the rehabilitation process begins in inpatient settings, the efficacy of treatment should continue to be assessed in the home setting via monitoring of the individual's physical activity level (e.g., shopping, joining an exercise program, doing home exercise, performing more household activities). Without reliable and valid baseline data on physical activity performance levels, it is difficult for rehabilitation professionals to monitor this important health-enhancing behavior when a person returns home.

Urgently needed in the field of physical activity and disability are measurement instruments that can capture physical performance levels with accuracy and subsequently identify an appropriate plan of action to ameliorate certain health conditions and keep people at a high enough functional level that they can continue performing various ADL or IADL independently. The overwhelming majority of assessment tools used in rehabilitation to measure functional performance in people with disabilities do not focus on physical activity or fitness as part of the recovery process and long-term maintenance of optimal health and well-being. With the growing emphasis on encouraging individuals with disabilities to become more physically active after rehabilitation and throughout the life span, measurement tools that can accurately assess physical activity behavior in people with physical, cognitive, and sensory disabilities will increase in importance.

Impact of the Environment on Health and Wellness

The *Healthy People 2010* chapter, "Disability and Secondary Conditions," suggests that the significantly lower rate of participation among people with disabilities may be related to environmental barriers, including architectural barriers, organizational policies and practices, discrimination, and social attitudes, and recommends that public health agencies begin to evaluate which environmental factors enhance or impede participation (U.S. Department of Health and Human Services, 2000). Common barriers to physical activity expressed by people with disabilities include inaccessible equipment and programs, transportation difficulties, limited income, unsafe neighborhoods, and a variety of personal barriers such as chronic health conditions and beliefs that exercise will not benefit one's health and may in fact have an adverse effect (Kinne, Patrick, & Maher, 1999; Rimmer, J. H., Riley, B., Wang, E., Rauworth, A., & Jurkowski, J., 2004).

There is an important need to better understand how the environment plays a role in health and functioning among people with disabilities. This can occur only after appropriate measurement instruments have been developed. These measurement instruments must be sensitive enough to facilitate change to existing environments (e.g., fitness and recreation centers). While most environments have general constructs related to accessibility that can apply *across* settings (e.g., built structure, equipment, program, professional), embedded within each of these constructs is an underlying set of items that are *setting specific* (e.g., items needed to access a community swimming pool will be different from items needed to determine the accessibility of a park, trail, or weight training facility). Identifying factors that make a particular environment accessible or inaccessible requires measurement instruments that address specific aspects of *that* environment.

Measurement instruments also must offer professionals and staff (e.g., fitness trainers, park rangers, recreation specialists, exercise instructors, architects, park district administrators, city planners and managers), as well people with disabilities and their caregivers, enough information to make the environment accessible in a cost-effective and meaningful way. While checklists have been developed to assess the accessibility of the built environment, they usually do not provide enough detail to isolate the *specific* modifications needed to allow successful participation by people with disabilities. It is important to begin to explore which environmental factors are key correlates of physical activity and to determine if these relationships are consistent across disability groups, settings, and locations. Instruments that can provide valid and reliable measures of the accessibility of these environments are the first step toward achieving this goal.

Given the critical need to increase physical activity participation among people with disabilities, an instrument has been developed to measure the accessibility of fitness and recreation settings for people with mobility limitations. Project AIMFREE (*Accessibility Instruments Measuring Fitness and Recreation Environments*) was a three-year research project funded by the Centers for Disease Control and Prevention to develop and validate a series of measures that can be used by people with mobility limitations and professionals (e.g., fitness and recreation center staff, owners of fitness centers, park district managers) to assess the accessibility of recreation and fitness facilities. The instruments address accessibility factors related to structure (building accessibility), program (accessibility of group classes), equipment (strength and cardiorespiratory machines), professional behavior (attitudes and knowledge), and systemic factors (policies and procedures of fitness and recreational facilities). The AIMFREE instruments have demonstrated good reliability and validity and can be used to assess the environmental accessibility of fitness centers, swimming areas, parks, and trails (Rimmer, J. H., Riley, B., Wang, E., & Rauworth, A. 2004).

Measurement Issues in Disability and Physical Activity

The diversity of disability from a demographic, environmental, and functional (physical and psychological) perspective makes measurement of physical activity a complex issue. Given the relatively small numbers of some disabilities such as spinal cord injury, multiple sclerosis or cerebral palsy, obtaining appropriate sample sizes with adequate statistical power for conducting physical activity research is often difficult to achieve. There are a host of measurement issues related to physical activity that must be addressed in persons with disabilities and several of these are described below.

Heterogeneity of Disability

The widely varying levels of health and function among people with physical, cognitive, and sensory disabilities makes it extremely difficult to develop assessment instruments that are broad enough to be used by individuals with varying levels of function yet specific enough to capture unique activity patterns. Many people with disabilities are unable to read regular print (e.g., those with blindness) or cannot read materials above a certain reading level (e.g., persons with cognitive or learning disabilities). Likewise, some individuals are unable to write (e.g., persons with quadriplegia) or have difficulty with comprehension (e.g., those with head injury, mental retardation). Instruments must therefore be developed in alternate formats (e.g., Braille, audio) and simultaneously assessed for their reliability and validity.

Another issue with physical activity surveillance is that many surveys are completed by key informants (e.g., caregiver, employee) and must be assessed for accuracy, since it is difficult for a respondent to know the entire spectrum of the person's physical activity behavior. This could introduce substantial error and limit the accuracy of the survey responses.

Progressive conditions such as multiple sclerosis, Alzheimer's disease, and rheumatoid arthritis may change frequently over time and differ substantially within the same disability group. This makes it difficult to assess a level of function related to a specific intervention, because the progressive nature of the disability may cause large fluctuations over the short and long term, with the time sequence for each condition varying substantially between and within various subgroups of disabilities.

Measuring Physical Activity for Improving Specific Health Outcomes

The small number of physical activity instruments that have been developed for disabled populations have been used only in short-term studies (e.g., 8-16 weeks). There are no longitudinal studies that have supported the utility of these instruments in accurately assessing the relationship between physical activity behaviors and specific health outcomes. It is unclear what *types* and *amounts* of physical activity are needed to achieve desired health benefits among various disability groups. Valid and reliable instruments are needed to capture critical information on the risk–benefit ratio of physical activity participation and improvement in health outcomes while examining the potential risk of injury from various modalities and volumes of physical activity.

There are also no published instruments that assess physical activity across a diverse target population (e.g., people with physical, cognitive, and sensory disabilities). This makes it extremely difficult to compare the impact of various health promotion and physical activity interventions on improving long-term health and function in diverse populations.

Physical Activity and Disability Surveillance

Tracking health and fitness status longitudinally is critical to understanding how exercise, and physical activity in general (e.g., household or employment) can affect various health factors in persons with disabilities. Although many physical activity surveys have been developed for the general population, few of these instruments have been specifically designed to capture the unique activity patterns of specific subgroups of disability (e.g., women, children, physical disability vs. cognitive disability). Tortolero and colleagues (Tortolero et al., 1999) concluded from their qualitative assessment, based on focus groups of minority women with disabilities, that there is a need to develop physical activity surveys that are more relevant for women, include well-defined and inoffensive terminology, and improve recall of unstructured and intermittent physical activity. Busy mothers attributed negative connotations to the use of leisure-time terminology and mentioned sports, exercise, and leisure the least frequently in reporting physical activity. The restriction of physical activity measures to these latter activities may account for reports that women get less physical activity than men. Activity measures also tend to be linear, assuming one activity at a time, when women often multitask or engage in several activities simultaneously.

Use of Self-Report Measures of Physical Activity

Researchers often use surveys to measure the physical activity profile of study participants, particularly in clinical trials and large epidemiologic studies. Whereas paper-and-pencil self-report measures provide a convenient and inexpensive means of assessing physical activity, the efficacy and sensitivity of these measures are complicated by a number of factors.

First, responders to self-report physical activity measures tend to be inaccurate in the recall of their past physical activity habits. Durante and Ainsworth (1996) contend that the large degree of unexplained variance in the report of physical activity is attributable to cognitive failures in recalling and reporting physical activity. People with cognitive disabilities (e.g., mental retardation, Alzheimer's) have a difficult time with recall, and therefore data may be inaccurate. Often, key informants (e.g., caregiver, residential worker) are used to obtain data but it is unknown how much the person knows about the child's or resident's physical activity behavior. Coyle et al. (2000) also noted that a substantial number of persons with disabilities engaged in lighter activities more frequently than in moderate or vigorous activities, and that these activities were often performed on an irregular basis, making them difficult to recall.

Second, many physical activity instruments have been validated on specific subgroups of the general population (e.g., children, adults). These instruments often do not capture very low levels of reported physical activity and contain questions that have little or no relationship to the lifestyles of many people with disabilities (e.g., "During the past month, did you participate in any physical activities such as jogging, calisthenics, or golf?"). One function of a survey instrument is to capture baseline measures so changes in behavior or performance can be tracked over time. Instruments that start at relatively high levels of physical activity than generally observed in many disabled populations (e.g., sport and recreational activities) will miss lower levels of physical activity (e.g., moving around, therapy, standing). The failure to capture very low levels of activity results in floor effects and sample data that are not normally distributed.

In response to the limited applicability of physical activity instruments to persons with disabilities, Rimmer and colleagues (2001) developed the *Physical Activity and Disability Survey (PADS)*. The PADS was designed as a semistructured interview in which respondents are asked about their physical activity behavior in the following domains: (1) exercise, (2) leisure-time physical activity (e.g., less structured physical activity), (3) indoor and outdoor household activity, (4) wheelchair use, (5) employment-related activity, and (6) physical therapy-related activity. In each of these areas, the respondent is asked if he or she participates in the given type of physical activity, and if yes, is asked more specific questions about the type, frequency, and duration of these activities. Despite efforts to make the PADS sensitive to low levels of physical activity, 22.9% of respondents reported no physical activity in any of the categories just mentioned (i.e., household activity, wheelchair ambulation, etc.). This is illustrated in figure 7.3 by the bar on the far left side, which represents respondents reporting no physical activity.

Conversely, instruments that focus solely on low levels of physical activity are also problematic. Such instruments limit the ability of researchers to make

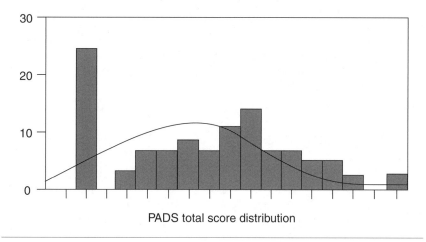

PADS total score distribution

FIGURE 7.3   Frequency distribution of the Physical Activity and Disability Survey total score.

comparisons between different disability groups or groups defined by levels of functional impairment. Ceiling effects that result from such measures may be as problematic as floor effects when one is evaluating the efficacy of health promotion or rehabilitation interventions.

Comparison of Self-Report and Objective Measures of Physical Activity

Objective measures, including pedometers, accelerometers, and ambulatory monitors, represent an alternative approach to measuring physical activity. These measurement tools have shown promise in their ability to predict energy expenditure and cardiorespiratory fitness compared to self-report instruments. The use of objective methods of physical activity measurement, however, is not without its limitations. Among individuals with mobility impairments, particularly those who use wheelchairs and similar assistive devices, the accelerometers, pedometers, and ambulatory monitoring devices fail to accurately capture the types of physical movement these individuals perform, since these instruments were designed to capture lower extremity movement.

A second limitation of objective measures is that their ability to predict energy expenditure may vary depending upon the nature of the activity being monitored. For example, two studies (Freedson, Melanson, & Sirard, 1998; Trost, 2001) showed that the Computer Science and Application (CSA) Inc. activity monitor successfully predicted energy expenditure (kcal/min) for treadmill walking. In contrast, Gordon, Heil, and Bauer (1999) observed that the CSA significantly overpredicted energy expenditure for mild, moderate, and vigorous levels of overground walking. Still other findings suggest that accelerometers may *underestimate* energy expenditure, particularly as workload during exercise increases. Accelerometers also appear to be less effective in predicting energy expenditure (peak $\dot{V}O_2$) for general lifestyle activity (r = 0.33) compared to energy expenditure for treadmill activity (r = 0.85). Finally, regression equations used to predict energy expenditure from objective measures may be specific to the population, activity, and type of measurement device used, limiting the generalizability of these regression equations to individuals with specific types of disabilities.

Functional Assessment of Physical Activity

Despite comments that physical activity may improve function among people with disabilities (Heath & Fentem, 1997; Rimmer, Braddock, & Pitetti, 1996), studies have not yet addressed the functional impact of physical activity among this population. Measures typically used to assess the effects of physical activity on people with disabilities are the more traditional physiological outcomes such as aerobic capacity, strength, body weight, and body fat. Unfortunately, indices that would more directly assess function-related outcomes have been underutilized in people with disabilities.

Several test batteries have been developed to assess these functional fitness improvements in older adults (Osness et al., 1996; Rikli & Jones, 1999). These tests generally include balance and strength measures, walking speeds and distances, stair climbing, and reaching tasks. Similar types of outcomes should also be assessed among people with disabilities. However, the test batteries used for older adults require the ability to stand and walk and may therefore be inappropriate for individuals with mobility impairments who have difficulty standing or who are unable to walk. Therefore, researchers should begin to identify fitness changes that can affect performance of functional activities by those with mobility impairments and develop tests of functional performance that are sensitive to detecting physical activity-related changes.

Conclusion

The relationship between physical functioning and physical activity among people with disabilities is a reciprocal one; physical functioning provides the individual with the capability to engage in physical activities, whereas physical activity helps to maintain and in some cases improve physical functioning. This reciprocal relationship, coupled with the high prevalence of physical inactivity among persons with disabilities, has profound implications for rehabilitation practice, especially in the evaluation of intermediate- and long-term outcomes of clinical practice.

For rehabilitation to play a role in the long-term maintenance and enhancement of physical functioning among persons with disabilities, monitoring of physical activity by both rehabilitation professionals and individuals with disabilities must be part of the recovery and maintenance process. The development of valid, reliable, and conveniently administered assessment tools to measure health and function among people with disabilities can serve the purpose of monitoring physical activity patterns at both the individual (clinical application) and population (public health surveillance) levels. We currently lack the ability to accurately assess the impact of physical activity behaviors on the health status and functioning level of people with disabilities. Assessment instruments that measure behaviors at the *person level,* as well as factors that increase or decrease participation at the *environmental level,* are needed for understanding the dynamics of disability interlaced with various levels of support in the environment.

Chapter 8

Measuring the Ever-Changing "Environments" for Physical Activity in Older Adults

James R. Morrow, Jr., PhD
Dale P. Mood, PhD

Physical activity is beneficial for all people, regardless of age. The evidence summarizing the relationship between physical activity and health is widely accepted and has resulted in numerous position statements about physical activity (e.g., U.S. Department of Health and Human Services, 1996; Thompson et al., 2003). Given that physical activity provides physical and psychological health benefits, focus should be directed toward intervention strategies that encourage individuals to initiate, adopt, and maintain healthy physical activity behaviors. Interventions should utilize effective adoption and maintenance strategies (e.g., health belief model, transtheoretical model) for all individuals. Yet many adults, particularly older adults, remain inactive (see figure 8.1).

Once the specific, individual readiness stage (e.g., Dishman, 1994; Marcus & Forsyth, 2003) has been identified, it is important to provide the specific, appropriate environment so that the likelihood of adoption and maintenance of desired behaviors is increased. For individuals to adopt healthy physical activity behaviors, the environment must be appropriate and conducive to physical activity. That is, environment is a key component to achieving healthy physical activity behaviors in all people.

Environment is a multidimensional concept. Environment can be viewed as structural, atmospheric, personal, psychosocial, and so on. Indeed, environment can denote surroundings, scenery, setting, situation, atmosphere, milieu, location, ecosystem, bionetwork, locale, scene, spot, position, and place, among others. Often, environment is considered only in the physical sense. This is rather shortsighted. Environment might be classified into four broad categories:

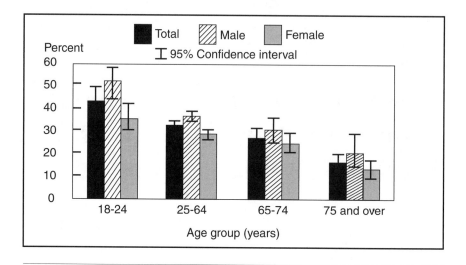

FIGURE 8.1   Percentage of adults aged 18 years and over who engaged in regular leisure-time physical activity: United States, 1997 to 2003.
www.cdc.gov/nchs/data/nhis/earlyrelease/200309_07.pdf.

(1) physical, (2) atmospheric, (3) psychosocial, and (4) political/advocacy. These broad categories are further delineated in table 8.1. The table lists examples of variables that constitute each of the environmental categories. It will serve researchers well to reflect on all of the measurement and research characteristics and environmental categories that must be considered when the ultimate goal is to influence physical activity behavior.

These four environmental categories affect all people engaging in physical activity, regardless of age. Consider the elementary-age child enrolled in a physical education class. The physical environment (gymnasium, playground, or classroom) will influence the amount and type of physical activity in which the child engages. Does the physical atmosphere (e.g., climate or altitude) encourage or present a barrier to physical activity? Likewise, the psychosocial environment affects what the child does. Are there too many children for the available space? Are classmates of differing ages and physical abilities? Does the child have a personal, social, or affective history that influences the amount of physical activity in which he or she engages (i.e., facilitates or hinders)? Do school administrators, teachers, and curricular decisions encourage physical activity?

While these four environments influence all people, the focus here is on how they affect the aging population (i.e., those above age 50). If, as suggested, each of these environments affects an individual's engagement in physical activity, it is important that we determine which aspects of these environments facilitate physical activity and which provide barriers. Also, we need to determine whether these environments and responses to them are generic for subgroups of the population or are idiosyncratic to the individual.

TABLE 8.1 Aspects of Physical, Atmospheric, Psychosocial, and Political/Advocacy Environments Related to Engaging in Physical Activity

Physical	Atmospheric	Psychosocial	Political/Advocacy
Availability	Altitude	Types of people involved	Awareness
Bicycle paths	Climate	Age	Federal
Building	Humidity	Attractiveness	International
Convenience	Lighting	Behavior determinants	Legislation
Direct observation	Temperature	Body type	Local
Ease of street crossing	Wet bulb globe temperature	Buildings	Policy
Frequency counts		Condition	State
Isolation factor		Connectivity	
Design		Culture	
Number of people involved		Population density	
Physical barriers		Distance	
Physical surroundings		Diversity	
Facility quality		Fear	
Perception		Gender	
Reading materials/Ability to read		Health	
Scenery		Isolation factor	
Self-report		Knowledge of health benefits	
Space		Knowledge of how to be physically active	
Speed limits		Loneliness	
Street patterns		Mind mapping of all "environments"	
Transportation		Mood	
Width of roads		Personal considerations	
Width of walkways		Safety	
		Self-efficacy	
		Social considerations	

It is suggested that much of the focus on environment has been on the physical environment. Much has been written about access, availability, safety, transportation, and so forth. A reader might inaccurately assume that creation of a quality physical environment will result in increased physical activity. Table 8.2 presents a list of recent journal issues related to changes in the physical environment that are expected to influence physical activity behaviors. While there is little doubt that the physical environment affects an individual's decision to engage or not to engage in physical activity, the multidimensional nature of environment must also be considered. It is shortsighted simply to provide a conducive physical environment, with the idea "If we build it [the physical environment], they will come [and be physically active]."

TABLE 8.2   Recent Publications Related to Environment or Physical Activity in Aging Populations

Publication and date	Subject
American Journal of Preventive Medicine — August 2002	Innovative approaches to understanding and influencing physical activity
American Journal of Public Health — September 2003	Community design
American Journal of Health Promotion — September/October 2003	Health-promoting community design
American Journal of Preventive Medicine — October 2003	Physical activity: preventing physical disablement in older adults
American Journal of Preventive Medicine — February 2005	Active Living Research

The other environmental dimensions also influence the decision to be physically active. Researchers must consider the interactive nature of the various environments and determine how they influence physical activity behaviors. In fact, all of these environments interact to influence the choice to be or not to be physically active.

The measurement and research issues are multiple when one considers the number of ever-changing environments. Some physical and atmospheric measures can be easily obtained with a measuring tape, frequency count, or thermometer. But measures in the psychosocial and political/advocacy environments are often difficult to obtain. In addition, the interactions among these environments (and the multiplicity of subdomains) and their influence on physical activity adoption are quite unwieldy. Figure 8.2 presents a model of how these various environments might influence the ultimate objective, that of increasing physical activity. Note that only the four major environments are presented in this figure. Each environment category contains multiple subdomains that could (and should) be measured. Identifying the direct and indirect influences of environment on physical activity behavior is clearly a multivariate problem.

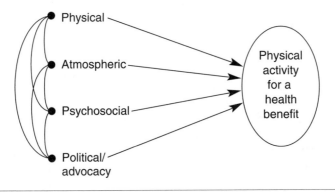

FIGURE 8.2   Schema of relations between environments and physical activity outcome.

Measurement Issues

As with all measurement issues, the key concerns are reliability, objectivity, validity, and relevance. These concerns subsume bias, accuracy, scale development, timeliness, population specificity, readability, readiness, action, and so on. The following list provides several examples of measurement outcomes and issues in assessing environments.

Sample Measurement Outcomes and Issues in Assessment of Environment

Affective domain
Assertiveness
Audits
Barriers
Bias
Colinearity/Independence
Counts
Dimensionality
Humidity
Influences
Internal consistency
Legislative actions
Length
Number
Objectivity
Perception
Personality
Psychological assessment

Qualitative assessment
Quantitative assessment
Radiation
Relevance
Reliability
Safety
Self-report
Sensitivity to change
Slope
Social assessment
Stability
Statistical modeling
Subscales
Temperature
Validity
Voting populations
Width
Willingness to engage

Statistical Analysis

Given the multiplicity of environments that influence physical activity, there is a broad range of statistical and research procedures that should be considered in assessment of environmental influences on physical activity. These include audits, frequencies, percentages, attitudinal questions, reliability coefficients (internal consistency and stability), multidimensionality considerations, relational statistics, and a priori model development. Researchers should not focus on counts or frequencies and the physical environment alone. The interactive effects of the various environmental categories must be taken into account.

Sample Instrumentation Issues

Assessment of several environments requires various types of measurement procedures and instruments. Among the environments to be assessed are the physical, atmospheric, psychosocial, and the political/advocacy.

The Physical Environment

The physical environment can be assessed in a variety of ways. Checklists and physical observations are appropriate. Audits of the physical environment are useful. Frequency counts of resources, facilities, and barriers can be conducted. See Saelens et al. (2003) for examples of multidimensional scales of neighborhood environments and Moudon and Lee (2003) for examples of environmental audits. These environment measures could also include considerations such as slope of trails, number of trails, width of sidewalks, and safety matters. Indeed, much of the literature on the influence of environment on physical activity behaviors focuses solely on the physical environment.

The Atmospheric Environment

The aged respond physiologically and psychologically differently to hot and cold environments than do young individuals. Thus, it is important to provide individuals with safe atmospheric environments for physical activity. Factors such as the wet bulb globe temperature (WBGT) can dramatically influence physical activity in the aging population. Initial fitness level is important in determining the type and amount of physical activity in which aging populations can engage (Kenney, 1993).

The Psychosocial Environment

The psychosocial environment is largely concerned with perception. Internal consistency, instrument dimensionality, freedom from self-report bias, and construct validation procedures underlie the development of sound psychosocial instrumentation. Thus, instruments need to be developed that provide a multidimensional approach to assessment of the affective domain. Older adults might respond differently than younger adults and children to aspects of the affective domain as it influences physical activity. For example, self-efficacy might be more important for older adults than younger individuals when people are choosing to engage in or maintain healthy physical activity behaviors (e.g., Clark, 1997). See also McAuley, Blissmer, Katula, et al. (2000) and McAuley, Blissmer, Marquez, et al. (2000).

The Political/Advocacy Environment

Measurement in the political/advocacy environment involves assessment of willingness, readiness, and abilities to influence decision makers as well as documentation of legislative authorization. Various organizations such as the American Association of Retired Persons, the President's Council on Physical Fitness and Sports, the Robert Wood Johnson Foundation, the U.S. Social Security Administration, the American College of Sports Medicine, the National Coalition for Promoting Physical Activity, and the American Alliance for Health, Physical Education, Recreation and Dance are attempting to influence older adults to become physically active through advocacy activities. Much of this work is subsumed under the National Blueprint: Increasing Physical Activity Among Adults Age 50 and Older, which currently consists of more than 50 support organizations (www.agingblueprint.org).

The Challenge

Measurement theorists, interventionists, and researchers must address the multidimensional nature of the environment and the ways in which the categories relate to physical activity in aging adults, with the ultimate goal of increasing physical activity to a level that provides a physical or psychological health benefit or both. Such action and influence will take more than investigating individuals' physical environments; research among the various environmental categories is necessary. Influencing anyone to be physically active, let alone an entire population, takes more than building a facility and having a sunny day, social support, and effective advocacy. All of these environments and their many subdomains are needed to potentially influence an individual to become physically active.

Most important, it may not be a generic change in these factors that influences the population to alter behaviors. More likely, it will be change in the environments on an individual level, because each person's decision to engage in physical activity is personal. Thus, influencing the individual unit (i.e., person) necessitates examining the highest-level order of statistical interaction. That is, a general model may not work well, but the N = 1 model considering the specific multidimensional aspects of the environment for the individual might be effective. Individuals could well be at different locations of the various subdomains of environments and be able to be influenced only at an individual level.

Future Research

The following are some research questions that address environments. What environments actually encourage people to engage in increased physical activity? Are these facilitation mechanisms generalized or are they idiosyncratic to individuals? What specific interactions among various environmental components result in increasing an individual's physical activity? Perhaps like determining "readiness for adoption," one needs to consider "readiness of the many environments" for one to become physically active.

Have researchers been shortsighted and tunnel-visioned when looking at environments that affect physical activity? Should they consider the multidimensional nature of the environment and how the various environmental categories interact to influence physically active lifestyle behaviors? These environments do not exist in isolation but rather in the milieu of one's life. Each person's perception of the various environments is different, resulting in great difficulty for influencing individual behaviors.

We can count sidewalks, parks, bicycle paths, and available space. While providing a suitable physical environment is a necessary condition to encourage increased physical activity, it is not in itself a sufficient condition. Unfortunately, just having these facilities available in safe environments does not necessarily mean that individuals will increase physical activity. We must consider all of the environmental dimensions that influence individuals when making lifestyle behavior choices. As with "mind mapping" or geographical information systems (GIS), it takes multiple measurements to pinpoint accuracy (or to influence behavior).

Only when we understand the wide range of environments influencing physical activity and the relations among these environments will we be able to influence physical activity. It is not an easy task to understand these relations, and it is most difficult to truly influence behaviors. It is suggested that this will take years of study. However, given the importance of physical activity in quality of life, the challenge before us is an exciting one, and one that offers rich rewards.

Chapter 9

Translating Research to Practice

Real-World Evaluation and Measurement Issues in Moving From Efficacy to Effectiveness Research

Marcia Ory, PhD, MPH
Diane Dowdy, PhD
Brigid Sanner
Robin Mockenhaupt, PhD, MPH
Laura Leviton, PhD
Russell Glasgow, PhD
Abby King, PhD
Cynthia Castro, PhD
Michele Guerra, MS, CHES
Sara Wilcox, PhD

The Evolving Research Base

The evidence supporting the linkages between physical activity and a myriad of health outcomes is now well accepted (U.S. Department of Health and

Human Services [USDHHS], 1996, 2003). The benefits of physical activity throughout the entire life course also have been increasingly noted (American College of Sports Medicine, 1998; Jette et al., 1999; Robert Wood Johnson Foundation, 2001). Despite aging stereotypes, we now know that it is never too late to increase physical activity—and always too soon to quit (Ory et al., 2003). Physical activity is seen as a leading health indicator (USDHHS, 2000) and is linked to the prevention and management of virtually every major chronic disease and disability (Agency for Healthcare Research and Quality and Centers for Disease Control, 2002; Bush, 2002).

Recent research has specified the physical activity–health relationship for different populations and begun to quantify just how much physical activity is necessary for which types of desired outcomes. Whereas the original focus was on vigorous activity, the value of moderate activity has been increasingly recognized. Similarly, there is a new emphasis on active living in addition to more traditional forms of "exercise" or other structured activity as through sport (National Institute on Aging, 2001; USDHHS, 1999).

In addition to epidemiological studies, there have now been numerous studies examining the determinants or correlates of physical activity—or its converse, physical inactivity and sedentary lifestyles—in later life (King, Rejeski, & Buchner, 1998; King et al., 2000; King, Bauman, & Calfas, 2002). While barriers and facilitators at the individual psychosocial level have received the most study, recent research reports are reflecting a multilevel approach in examining influences using a social ecological framework (McLeroy et al., 1988; Sallis & Owen, 2002; Smedley & Syme, 2000; Stokols, 1996). Intrapersonal, interpersonal, organizational, community, and policy factors are all seen as potential influences and targets of intervention.

This growing body of evidence has led to myriad recommendations for increased physical activity. There is now greater specification of a common health message—Americans are called upon to engage in 30 or more minutes of moderate or more vigorous physical activity on most (preferably all) days of the week (American Association of Retired Persons, 2003; Pate et al., 1995). Although the newer emphasis on moderate-intensity activities may make these goals more attainable by larger numbers of older people than the previous recommendations for vigorous activities, there is also a greater challenge in measuring lower intensity levels of activity reliably.

Despite more inclusive definitions of leisure-time activity, national studies still indicate that the majority of adults, and especially older adults, fail to meet these recommendations. For example, two-thirds of older adults do not engage in any physical activity at all (Centers for Disease Control and Prevention, 2004). Instead of improvements in physical activity levels, trend data show that activity rates have not improved over the past decade, and, alarmingly, associated conditions such as obesity levels are actually rising (Center on an Aging Society, 2004).

Principles of Behavior Change Research

Given the well-documented costs of sedentary lifestyles (Booth & Chakravarthy, 2002), there is renewed emphasis on testing the efficacy of theory-based interventions to encourage adults 50 and older to initiate and maintain more active lifestyles. Several principles of behavior change research are emerging from this perspective (Ory, Jordan, & Bazzarre, 2002). These principles help inform choices related to theoretical application, study design, recruitment and retention activities, treatment implementation or fidelity activities, assessment issues, and targeted outcomes.

- Behavior change occurs within broader sociocultural contexts, whereby initiating and maintaining healthy behaviors are best accomplished through identifying multiple opportunities for intervention in people's physical and social environments.
- Interventionists need to better understand mechanisms and moderators of change, and target these variables.
- The goals of healthy behavior change are more successful when they are concrete and are arrived at through collaboration between the interventionist and the target group of interest (e.g., the individual participant, worksite population, community).
- The goals of behavior change should be conceptualized as moving targets based on initial performance levels, rather than absolutes.
- The adoption and maintenance of healthy behaviors require basic skills that can be taught, practiced, and used effectively in "real-world" contexts.
- Adherence to and maintenance of healthy behaviors typically benefit from planned "boosters" (i.e., repeating or enhancing key parts of the initial intervention).
- Sustained change in target populations requires the development of a variety of interventions aimed, either singly or in combination, at multiple levels of influence.

Knowledge of these principles can help direct selection of variables to be measured and the optimal research designs to be applied (e.g., the importance of longitudinal designs to assess change over time with attention to maintenance or relapse issues). While there is a relatively strong measurement history for psychosocial mediators and behavioral physical activity outcomes, for other critical issues there is a dearth of rigorous investigation. For example, what are the best measures for contextual factors, such as neighborhood environmental factors or social cohesion? How does one measure capacity for sustainability and indicate extent of program adaptability? Additionally, there is special difficulty in assessing efforts at transbehavioral change—that is, the ability to measure behavioral change across multiple lifestyle practices.

Behavioral Change Consortium

The Behavioral Change Consortium (BCC), a collaborative partnership of 15 NIH-funded studies designed to test multiple behavior/multiple intervention strategies, is an excellent example of cutting-edge behavioral intervention research (Ory, Jordan, & Bazzarre, 2002). The BCC studies were designed to further knowledge on behavioral processes of change, and treatment fidelity was a major emphasis. While most were affiliated with academic research settings, a few involved tests of evidence-based interventions in real-world settings such as schools, churches, and health care environments.

Behavioral Change Consortium investigators advanced intervention research knowledge in several cross-cutting areas: (a) recruitment and retention of hard-to-reach populations, (b) strategies for enhancing treatment fidelity, (c) measurement and assessment across multiple lifestyle behaviors, and (d) exploration of ways to maximize reach and translation. Eleven of the BCC projects invited study participants to engage in physical activity/exercise interventions, with three being specifically focused on interventions for older, and even frail, populations.

The Physical Activity work group had lively discussions about the best measurement scales and designed substudies to assess the validity of different measurement strategies in the three aging studies. A major topic of discussion was whether or not there could be an absolute criterion of success (e.g., meeting the Surgeon General's recommendations) or, given the full range of older adults studied, including frail elders after a hip fracture, if it was more appropriate to talk in terms of relative improvement from an individual's baseline status.

Given the limited time frame for study, these investigations focused primarily on behavioral initiation versus long-term maintenance. Comprehensive measurement batteries were also typical, with extensive batteries for both mediational and outcome variables. Targeting and measuring multiple health-related lifestyle behaviors (e.g., smoking, nutrition, exercise) were especially challenging. Since most physical activity projects also measured at least one other lifestyle behavior (typically nutrition), the difficult issue of transbehavioral assessment was raised, but no adequate solution was found within the BCC. As is typical of efficacy-based research, these studies were randomized clinical trials with an emphasis on maximizing internal validity. Investigators in the Research and Translation group began to explore external validity issues, mapping out a RE-AIM framework for program implementation and evaluation (Glasgow, Vogt, & Boles, 1999). RE-AIM is a systematic way for researchers, practitioners, and policy decision makers to evaluate health behavior interventions. Ultimately, it can be used to estimate the potential impact of interventions at the population level. RE-AIM (Reach, Efficacy, Adoption, Implementation, and Maintenance) is managed by the Kansas State University Research and Extension Community Health Institute. Interactive materials are available on the RE-AIM Web site (www.re-aim.org).

Going Beyond the Behavior Change Consortium Experience

The BCC experience highlighted the need for additional research efforts to better address translational research issues. Recruitment and retention are not "noise" in the system, but important areas of research for understanding how to expand reach and maintain intervention effects. New research endeavors are being called for that reintroduce old concepts into today's research settings. Diffusion of innovation concepts (Oldenburg & Parcel, 2002; Rogers, 1995) is being brought to the forefront with the current emphasis on research dissemination into community settings. Similarly, sustainability of innovative programs is now a key issue in research translation (Evashwick & Ory, 2003; Weiss et al., 2002). Additionally, new models are being generated in efforts to emphasize the generalizability of research in real-world settings, with the RE-AIM conceptual framework as one key example for planning as well as evaluating public health impacts of intervention efforts (Glasgow, Vogt, & Boles, 1999).

Active for Life®: Research to Practice

Research-based interventions are often successful in research-based settings, but typically are never tested in real-world settings or may fail to translate effectively. By the same token, and due in large part to limited resources and lack of research-trained staff, community-developed interventions are often not evaluated and thus their effectiveness is never demonstrated. In 2000, the Robert Wood Johnson Foundation (RWJF) sponsored a conference of 48 organizations with expertise in health, social and behavioral sciences, epidemiology, gerontology/geriatrics, clinical science, public policy, marketing, medical systems, community organizations, and environmental issues. The outcome was reported in a document called "National Blueprint for Increasing Physical Activity Among Adults Age 50 and Older" (RWJF, 2001). The "Blueprint" outlined both the challenges and broad strategies related to increasing physical activity among midlife and older adults. The *Active for Life* initiative addresses a number of these challenges.

Building on two decades of behavior intervention research for adults in the middle and later years, *Active for Life* was designed to take research-based programs into community settings, provide structured social marketing support, and conduct independent evaluation to measure effectiveness. The knowledge gained will be disseminated to professionals in the health, wellness, fitness, and aging fields.

The overarching goals of the *Active for Life* program are to learn how to deliver research-derived physical activity programs to large numbers of midlife and older adults, and to sustain such programs through existing community institutions, including, but not limited to, health, social, community, aging, religious, or recreation centers and agencies. *Active for Life* is a four-year initiative supported by RWJF that seeks to increase the number of American adults age 50 and older who engage in regular physical activity (at least 30 min a day on most days). This initiative tests the added utility of a multipronged strategy that examines tailored interventions based on the diverse social, behavioral, environmental, and functional circumstances of midlife and older adults. The *Active for Life* National Program Office, housed at the School of Rural Public Health at The Texas A&M University System Health Science Center, is one of several programs in the RWJF Active Living Consortium.

In addition to national program office oversight, the University of South Carolina Prevention Research Center is evaluating the program. Information from a two-phase evaluation will help the physical activity and aging fields understand how the intervention models work in real-world settings, as well as the effectiveness of the models in encouraging people age 50 and older to initiate and maintain physical activity. The field of public health and successful aging can benefit from knowing how these research-based models may need to be adapted when implemented in the community, and which subgroups of the age 50-plus population they are most effective for.

Good science is rarely the object in efforts to disseminate evidence-based prevention models, but this is a vital and neglected feature of prevention research. The answers bear on practicality, quality control, and external validity. The evaluation team will be conducting a detailed process and outcome evaluation. In the process evaluation, we hope to learn the ways in which the two selected program models must be adapted to be acceptable to community organizations and intended constituents, and then determine whether the adaptation is consistent with the core elements of the original program. If the adaptation is an acceptable variation on the original, it should be added to the knowledge base about the model and shared with others who are struggling to translate the research into practice. If the adaptation is *not* an acceptable variation on the original, that needs to be shared as well, for quality control. The primary reason to conduct outcome evaluation at the new sites, *once the pilot phase is over*, is to understand the extent to which outcomes are similar in the new adaptations, and for whom.

The heart of the initiative is an $8.7 million grant program that supports nine community-based organizations (with multiple sites) for a four-year period to test the effectiveness of two promising interventions to promote physical activity in the general population of midlife and older persons at health risk because of their sedentary lifestyles. Several criteria were utilized to select two interventions for further testing. The program needed to (1) be successful in increasing physical activity in the 50-plus population; (2) be a behaviorally based intervention that explicitly incorporated the principles of behavioral change research (i.e., be more than an exercise training program alone); (3) be a manualized program, that is, one for which program materials and facilitator's guide had been developed; and (4) be tested previously in multiple settings with a variety of participants.

The following are the two interventions that were selected:

• *Active Living Every Day (ALED)* is a comprehensive group-based behavior change program developed by behavioral scientists and interventionists at the Cooper Institute in Dallas, Texas. Active Living Every Day emphasizes a lifestyle approach to physical activity and encourages people to find ways to realistically fit physical activity into their daily lives. This program is based on extensive research that shows that teaching people lifestyle skills such as realistic goal setting, identifying and addressing barriers to physical activity, and developing social support systems will help people become and stay physically active.

• *Active Choices (AC)* is a comprehensive individually based program grounded in 20 years of systematic research and evaluation by public health researchers and community intervention specialists at the Stanford Prevention Research Center. It is an individually tailored, telephone-supervised program that provides instruction, feedback, and support to participants. It offers people the opportunity to choose when and where they undertake their exercise, while providing an ongoing level of physical activity advice and support that many people desire.

Another major element of the *Active for Life* program involves technical assistance to the grantees on the RE-AIM framework, developed to provide public health and community settings with a systematic way to approach health behavior promotion planning, design, and evaluation.

Lessons Learned From the Selection and Early Implementation Phase of Active for Life

We have already learned many lessons in our first years of this endeavor. The following discussion represents our initial general impressions. We will be studying these issues in greater depth over the course of this project.

Feedback in Pre-Awardee Phase

Paralleling a National Council on Aging (NCOA) survey of the aging services network (NCOA, 2001), we learned that there are a substantial number of ongoing physical activity programs in communities serving older adults, but that most programs are group-based "exercise training" versus behaviorally based programs. Surprisingly, there was more of a life span approach, that is, inclusion of 50-plus people in traditional aging networks, than we expected. Most applicant organizations were working as part of a coalition of community organizations to increase physical activity rather than trying to increase activity solely within their own organizational settings. Coalitions are a useful way of combining reach and resources, but there needs to be clear lines of communication across multiple partners. Effective leadership requires a lead partner that can direct how emergent issues and problems will be handled in a timely fashion.

Community organizations are eager to be involved in research and evaluation endeavors that bring in service dollars. We had approximately 500 letters of intent, and other related physical activity funding opportunities also had exceptionally high application rates. But ultimate success depends on the ability to bridge research and practice perspectives. We selected grantee sites based on their potential for recruiting the designated number of enrollees in both the pilot and implementation phases of the study, their ability to implement evidenced-based programs, their willingness to contribute to the process and outcome evaluation, and the likelihood that selected grantees could sustain programs over time and serve as a model for disseminating research lessons learned to others.

Challenges in Translating Research to Practice Communities want—and deserve—a say in basic research designs and evaluation methodologies. Randomized controlled designs may be the gold standard for academic researchers, but they are not often acceptable in community projects where emphasis is on serving *all* constituents. It is often hard to establish traditional control groups in field settings, since communities feel that everyone deserves something for participating in the study. Similarly, individuals may be less motivated to come back for follow-up assessments if they don't see the immediate benefit for them. Moreover, the typical university-based institutional review board paperwork requirements and increasingly legalistic stance often seem onerous to community-based organizations devoted to service delivery issues. That having been said, communities are very concerned about screening and safety protocols in terms of providing safe services for their clients and avoiding liability issues, as well as about sensitivities in approaching their clientele.

Measurement batteries need to be substantially truncated—and culturally acceptable. This may be seen as problematic by researchers who think in terms of comprehensive multidimensional assessments that have better psychometric properties. In a collaborative vein, we made some fundamental alterations based on feedback from community groups (e.g., simplification of screening assessment forms). The key is to try to reach a compromise that is seen as acceptable for the researcher ("good enough") and yet "do-able" and useful for the community organizations.

Communities know their current service constituents well, and have an easier time accessing them than external researchers would. Researchers can help communities with strategies for reaching out to new populations and going beyond current comfort levels in assessments. However, an essential ingredient is community trust; after this is established, measurements that initially seem threatening may be able to be assessed with little reluctance. For example, trust will facilitate the collection of basic demographic information about ethnic characteristics that may initially be interpreted as sensitive data.

Because of logistical and cost issues, we had to rethink the requirement that everyone would get a physical functional assessment. However, several groups took this assessment upon themselves since they saw a functional assessment as valuable feedback information for making the case for sustainability in their own settings. Our lesson learned was that community partners may be willing to "buy in" to more extensive assessments if they see the value of such measurements for their constituency as well as their organization.

Researchers need to think in terms of creating tool kits and templates that can be widely used in community settings and are adjuncts to recruitment and program implementation. Especially needed are tools that (a) help communities assess whom they are actually reaching and (b) help community organizations extend beyond the community segments they typically reach to those who may be underserved but may significantly benefit. Some approaches may result in productive yields but may not push programs beyond "business as usual" with their current clientele. For optimum use and effectiveness, these tool kits should emphasize principles and guidelines, not mandated lockstep protocols.

Behavioral scientists and program developers need to realize that in the field, programs are often implemented in ways that depart significantly from the theory-based interventions as originally designed. The question is how much adaptation a program can incorporate and still adhere to essential components. There is always a delicate balance between fidelity and adaptation-reinvention to fit particular settings (Rogers, 1995). Communities can tell researchers what works and doesn't work in their settings. Testing an unacceptable intervention is not really a good test of program reach and effectiveness. It is critical to assess what adaptations are being made—and the rationale for such adaptations—and whether essential program elements are actually implementable in field settings. This can help guide the next generation of applied intervention strategies. A comprehensive process evaluation can be quite useful in documenting these adaptations, reasons for the adaptations, and impact on outcomes.

Although the programs were not initially envisioned as community participatory research, grantee sites from the beginning wanted to make changes to the programs. Some of those changes, such as adaptations in the layout and font size of program materials to improve readability, were obviously helpful. Others, such as modification of some of the physical activity tracking forms, made sense; but determining how to change the forms proved to be complex. The forms needed to be adapted in a way that (1) made them easier to use for the participants, (2) captured the information that was important to the evaluation team, and (3) provided the educational and behavioral benefits to the participants that the program developers intended. Discussions among the grantee sites, program developers, evaluation team, and grant coordinators have helped us understand the perspectives of each group. Through these discussions we recognized the need for an ongoing review process. This enables grantee sites to adapt programs relatively quickly, while assessing the potential effects of the proposed adaptations. All parties have a voice in the process. This should help ensure that future adaptations will be collaborations between the grantee sites and program developers, with input from the grant coordinators and evaluation team when necessary.

While potential for sustainability was a key selection factor, in reality this issue is quite difficult in tight fiscal times, and it is difficult to promise sustainability without knowing what the eventual program results will be. Therefore, it is especially important for community grantees to build in sustainability plans from the very beginning. Communities and researchers have different definitions of success—and meeting one's organizational criteria of success may be more strongly related to sustainability than a researcher's predetermined measures. For example, in some communities, measures of success were not data based but instead were practical things such as individual success stories that could be shared, or the ability to retain lapsed clients. For other organizations, the impact of the program on health care costs and health care utilization may be critical for sustainability.

One area affecting sustainability that has become evident is the ability for the grantees to train their own intervention facilitators. The program developers trained the first group of facilitators. This approach has advantages: it provides an opportunity for program developers and implementers to meet, and it maximizes the chances that the grantees will implement the programs effectively. On the other hand, this approach is costly and, as grantee sites need to train larger numbers of facilitators due to program growth or staff turnover, difficult to sustain. The grantee sites, program developers, and grant coordinators are currently discussing how to develop a train-the-trainer program that protects the integrity and effectiveness of the programs while providing the grantee sites with the ability to sustain and grow the programs over time.

Key Measurement/Methods Questions

We are learning that there are no set protocols for translating research to practice. But there are certain principles of community participatory research that serve as road maps for this endeavor. Our fundamental stance is that this is a two-way process—whereby community partners and researchers will learn from each other. Additionally, both quantitative and qualitative research approaches can be useful. Along our travels we have identified several thorny design and measurement issues that we would like to bring forward for discussion among experts in the field. These include topics such as the following:

• How can one best measure reach of the program if there is not a clear denominator for comparing characteristics of the population reached with those of people who could be potentially reached? What tools are there to help community practitioners assess their success in reaching new and diverse populations?

• What is the best comparison group and design if the community insists that everyone get the service or intervention? Are wait list methodologies feasible if the waiting period is more than six months? In what ways can existing secondary data be employed for comparison purposes (e.g., state Behavioral Risk Factor Surveillance System data)? One alternative to the wait list or assessment-only control group that has been increasingly applied in the behavioral research field is the use of an attention control group that targets a different, though similarly attractive, health behavior for intervention (e.g., dietary change, stress management).

• How can we develop more community practice-friendly measures that can be incorporated into community organizations' daily service routines? We believe this is essential for continued assessment after the funded research period ends. For example, most self-reported physical assessment batteries are 10 items or more, especially when there is a need to differentiate light-, moderate-, and high-intensity activities. Often what is really needed is a one- or two-item assessment that is valid and sensitive to change, although realistically such instrumentation presents challenges for both researchers and community populations.

• Should outcomes be conceptualized in terms of percentage improvements or achievement of some criterion standard? When including adults with a wide range of functional abilities, one needs to acknowledge that some participants may never reach the designated criterion, yet these "suboptimal" changes may translate into meaningful impacts on the participants' quality of life and functioning.

• How can multiple interacting outcomes be tracked? There is often not one simple primary outcome; rather there are complex, interrelated factors and different priorities placed on the various potential outcome indicators of success. For example, in coalitions, some partners may focus on increases in physical activity, while others may focus on improvement in functional outcomes, quality of life, or cost-effectiveness indicators.

• How can subtle changes (e.g., small increases in physical activity levels) be tracked over time? Do interventions have an immediate decay effect after the program ends, or a lagged effect, and what are the best designs for determining both short- and long-term outcomes?

• How can we put outcomes in clinically meaningful terms—going beyond statistical significance? It is also challenging to find clinically meaningful outcomes in the typically short time frame for study (e.g., what changes are reasonable to expect in six months or a year?).

• How does one assess the impacts of changing interventions, where innovations in delivery are part of the overall test? And how can one assess the impact of multiple intervention strategies, when some of our behavioral interventions are embedded in larger intervention strategies (e.g., overall community development approaches)?

• How does one go about analyzing and interpreting findings from community-based studies that are not randomized control trials? In particular, how do we deal with unit of analysis issues when we have nine main grantee settings but multiple sites within each setting?

- How can we encourage more research funding for maintenance of behaviors? Is it ideal to assume measurement research will be embedded in substantive topics, or is it necessary to have separate funding streams for methodological and measurement research?

Furthering Translational Research

The promotion of translational research requires careful attention to the types of conceptual and methodological issues raised in this chapter. There is also a need for a basic paradigm shift from academic research endeavors to campus–community partnerships, and infrastructure support to assist in these efforts. We lament the lack of (or perceived lack of) mechanisms currently available to help shift empirical research into the hands of practitioners. Researchers generally are not trained, encouraged, or "incentivized" by their institutions or grant funding agencies to take time away from their randomized controlled trials and publishing to turn their research into products for dissemination. Without special mechanisms or initiatives (e.g., the *Active for Life* initiative), it's a rare investigative team that is able to turn their research findings into translatable and sustainable programs within the time and fiscal constraints of the original grant award. We strongly urge funding agencies and review bodies to build effectiveness research and dissemination efforts into their research expectations and funding criteria. Research translation, dissemination, and sustainability should be the ultimate objective of intervention research, above and beyond academic publications of the major intervention outcomes.

By the same token, we need to continue to work to bring innovative and apparently effective community programs to the attention of the research community, as a means of building bridges between research and practice in the reverse direction. Through such work, collaborations can be built to get community programs already in place more fully evaluated and get them disseminated through scientific as well as practice channels. This type of "two-pronged" approach is essential if we are to realize significant gains in physical activity at the population level.

Chapter 10

Qi, Aging, and Measurement

History, Mystery, and Controversy

Weimo Zhu, PhD

The astounding demonstration of Qi-gong by Master Chen, in which he moved a person using his Qi without physically touching the person, introduced a completely new, unknown phenomenon to many Western scientists at the M&E Symposium. As expected, many of the scientists were skeptical about what they observed, and many interesting questions were raised. Is this really a human ability we do not know about, or is it just a simple trick? Would a person with such ability have better health and a longer life? Perhaps the most important question for researchers is whether the ability can be measured or quantified. All these questions or suspicions are logical, which is often the case with thought about Qi-gong in China, where Qi-gong was born and developed. The purpose of this chapter is to provide a brief introduction to Qi-gong, including what it is, its history, and the mystery and controversy surrounding it. The relationship of Qi to health and aging will also be described. Finally, the measurement challenge and future directions in Qi-gong research will be briefly discussed.

What Are Qi and Qi-Gong?

The ancients in China came to the conclusion thousands of years ago that Qi (life force, or vital energy) is the essence of all things on earth, and that the entire universe is a movement or mutation of Qi. Therefore Qi is within the human body. In fact, the entire system of traditional Chinese medicine is founded on the concept of Qi. For example, Chinese herbal medicine considers whether a given herb is Yin or Yang in its energy (note that in Chinese, Yin means female and negative principle, and Yang means male and positive principle), whether it is a tonic for deficiencies in Qi or whether it clears and creates an outflow of Qi, what organ it affects, and so on. "Gong," in Chinese, means practice or training. Therefore, the two characters in "Qi-gong" literally mean "Qi training" or "working with Qi." (Note: "Qi-gong" is pronounced "chi kung" in English.) The gentle movements, easy postures, and simple meditation techniques bring about a balanced energy flow, the end result being optimum health of the body and mind. It should be pointed out that although the word "Qi" has been used in the Chinese literature for thousands of years, "Qi-gong" is a rather new term that was created in the 1950s. In ancient times, the term was "Dao Yin," which means directing the flow of Qi. The contents of, or ways to practice, Qi-gong are varied, but they usually involve three types of regulation: of body (posture), of respiration, and of mind. Self-massage and movements of the limbs are also parts of Qi-gong routines.

A Brief History of Qi-Gong

Qi-gong has a rich history of more than three thousand years in China. For a full understanding of Qi-gong, a brief review of the history of its development may be helpful.

Ancient Times

In ancient China, Qi-gong was used to improve one's health and cure disease. The earliest records of Qi-gong can be found in the "Jin Wen" (writing on bronzes) of the Zhou Dynasty (c. 1100-221 B.C.) (Li, 1985). During the Spring and Autumn and Warring States periods (770-221 B.C.), Qi-gong developed rapidly. In the famous book *Yi-Ching* (*The Book of Changes*), the ability of reproduction, internal energy, and the mind were regarded as the three treasures in the human body. Dao Yin, the ancient healing technique that combines regulated breathing with body movements, was recorded in the book *Neijing*, the "bible" of traditional Chinese medicine. During the later period of the Eastern Han Dynasty (22-220 A.D.), the physician Tuo Hua did a thorough study of the traditional theories and practice of Dao Yin and developed a set of Dao Yin exercises called Wuqinxi (pronounced "woo chin see"). Literally translated "five-animal play," Wuqinxi consisted of simulated movements of the tiger, deer, bear, ape, and bird. Bringing out to the full the characteristic features of Dao Yin, the Wuqinxi helped to limber up the muscles and joints and proved valuable in disease prevention and treatment. Quickly it became one of the most popular forms of Qi-gong exercise.

Perhaps the earliest written record of Qi-gong is the Dao Yin diagrams, discovered in 1972 and 1973 in Changsha, China. Changsha, the capital of Hunna Province, was a well-known Chinese city during the Spring and Autumn and Warring States periods (770-221 B.C.). Between 1972 and 1973, two Han tombs were excavated in Mawangdui, the eastern corner of Changsha. In one of the tombs (Tomb No. 3), which belonged to a general of the Western Han Dynasty (206 B.C.-24 A.D.), large quantities of weapons and books on medicine were found. Among them was a silk scroll on which more than 40 human figures in different postures were outlined in black and painted in color. While some of the figures and Chinese characters are barely legible, it is believed that they are a series of Dao Yin diagrams belonging to the early Western Han Dynasty. Figure 10.1 shows four drawings that appear in the scroll. At the top left of figure 10.1, for example, a person stands with the knees slightly bent and hands slightly extended outward, a posture found often in today's Qi-gong exercises. At the bottom left of figure 10.1 is a person bending forward at the waist, with the

FIGURE 10.1 Dao Yin illustrations from the Western Han Dynasty in China (206 B.C.-24 A.D.).

palms touching the floor and the chin drawn up as much as possible—another popular posture also found in modern Qi-gong exercises. These figures testify to the rich experiences gained by the Chinese people through age-long struggles against ailments and illnesses of one kind or another.

Period of Integration With Martial Arts and Religion

After the Eastern Han Dynasty (25-220 A.D.), Buddhism was introduced into China. Yaga, an ancient exercise form in India, was not known in China at the time. Eventually, Dao Yin and Yaga were integrated, and new theories and practices of Qi-gong were developed. It is said that during the Southern and Northern Dynasties (420-589 A.D.), an eminent Indian monk came to China and established Zen (Ch'an) Buddhism in Henan's Shao Lin Monastery. Integrating Qi-gong with martial arts (known as "Wu Shu" in Chinese or "Gong/Kung Fu" to many Westerners), he evolved a set of exercises to help keep fit and also serve as a form of self-defense. Because Shao Lin's monks had saved the life of one of the emperors of the Tang Dynasty (618-907 A.D.) after an accident, the Shao Lin temple and its martial arts were promoted and supported during the period of the Tang Dynasty. Gradually the exercise and self-defense routine became the foundation of Shao Lin martial arts, known as Shao Lin Gong Fu to the world today.

Along with the creation of other martial arts schools and religions, many other Qi-gong routines were also developed during this period. Tai chi, a familiar Chinese exercise, is a good example of such a development. Based on the Tao, tai chi chuan [chuan means fist in Chinese], as well as its Qi component, was a martial art when invented. Although it is used mainly for health promotion now, there is still a very small group of people who practice tai chi for the purpose of self-defense. It is said that once there were more than 200 styles and schools of Qi-gong (Yang, 1992; Zhang & Sun, 1985). The drawback in development during the later part of this period is that, because of its martial arts function, teaching Qi-gong was conducted in a controlled way, that is, only to selected students and family members, so as to keep it secret.

Modern Times

With the overthrow of the Ching Dynasty in 1911, China entered its modern period and the influence of Western culture began, including modern medicine, physical education, fitness concepts, and the education system. Qi-gong education became more public, although it was not integrated formally into school education. In 1949, the People's Republic of China was founded. During this period, Qi-gong masters applied Qi-gong successfully in rehabilitation. Because of its effectiveness and low cost, Qi-gong was promoted by the Chinese government. In fact, the term Qi-gong was officially introduced to the general public during this period. The good times for Qi-gong, however, did not last long. From 1966 until 1976, China underwent its Cultural Revolution, a culture disaster indeed. During that period, because of its association with religion, practicing Qi-gong was forbidden and Qi-gong literature was destroyed. The value of Qi-gong was re-recognized after the Cultural Revolution. Practice of Qi-gong resumed, and related literature was published openly again. Between 1980 and 1995, Qi-gong experienced perhaps its best period in China. Many forgotten routines were reintroduced, and millions of people started practicing Qi-gong. More importantly, scientists from multiple disciplines were working together to try to understand the true nature of Qi and Qi-gong.

Qi-gong then suffered a huge blow as a result of Fa-Lun Gong and similar practice groups. As public demand increased, some leaders of Qi-gong schools exaggerated the effect of Qi-gong, as practiced either by themselves or by their followers, and provided misleading medical advice and information to their followers. Others mixed in superstitious belief to meet their followers' psychological needs. When the media began to question such practices, the Chinese government started to regulate selected Qi-gong groups. Protests took place, a rare occurrence in China. Because overseas political parties directly organized some of the protests, the situation became even more complex. Because of this series of events, the Chinese government cracked down on Fa-Lun Gong, the major group involved in the protests. Unfortunately, many good Qi-gong practice groups were also limited during the crackdown, and the general public was frightened away from practicing Qi-gong. Although the Chinese government has recently introduced four ancient Qi-gong routines to try to reduce tensions and meet the general public's practice needs, the interest in Qi-gong practice and research has remained low.

Qi-Gong and the World

Most people outside of China learned of Qi-gong from "Gong-Fu" (i.e., martial arts moves, in which Qi and its applications are often exaggerated). Qi and Chinese medicine did not gain recognition by the Western scientific community until President Nixon's historical 1972 visit to China. On Nixon's trip, journalists were amazed to observe major operations being performed on patients without the use of anesthetics. Instead, wide-awake patients were operated on with only acupuncture needles inserted into them to control pain. During the trip, James Reston, a famous *New York Times* columnist, developed appendicitis. The Chinese proposed surgery for his appendectomy using acupuncture anesthesia. His postoperative pain after appendectomy treatment was relieved by acupuncture. Reston later wrote convincing stories on its effectiveness.

In fact, Qi, instead of needles, was also used for similar purposes during the same period in China. Gradually, practice-based Chinese medicine and the Qi concept, as in acupuncture, herbal medicine, and tai chi, have been accepted to some degree by the mainstream of Western medicine. To meet the general public's interest and need, the U.S. government has made a considerable effort to improve research and education on so-called complementary and alternative medicine. Congress established the Office of Alternative Medicine in 1992 and the National Center for Complementary and Alternative Medicine (NCCAM) in 1999. Funding for both the Office and Center has increased greatly during the past decade or so, from $2.0 million in 1992 to $123.1 million in 2005. Qi and Qi-gong are, however, still relatively new to the Western world even with the interest in and push to investigate such concepts.

Qi-Gong Schools and Classification

For a better understanding of Qi-gong, a few words about Qi-gong schools and routines may be helpful. According to the literature (e.g., Lin, 1987; Zhang & Sun, 1985; Yang, 1992), there are more than 2,000 different styles or routines of Qi-gong. Because of their close ties with religion and martial arts, each school usually has its own specific training method, which is often not available to the general public. Several criteria have been used for Qi-gong classification (Liu, Zhang, & Liu, 1992). According to the school, Qi-gong can be classified into five groups: medical, Confucian, Buddhist, Taoist, and Wushu (i.e., Chinese martial art). According to the way in which Qi is practiced and used, Qi-gong routines can be classified also as internal or external. Internal Qi-gong can be further classified as (a) stillness meditation, in which little body/posture movement is included and the focus is on breath and internal Qi flow; or (b) moving meditation, in which general, slow, smooth movements are included (e.g., tai chi). For external Qi-gong, Qi-gong healers, who usually have practiced for many years and have the ability to control their Qi for both internal healing (of self) and external healing (of others), use their Qi to help their clients or subjects with their weak or blocked Qi. This kind of Qi-gong is known as "energy therapy," as classified by NCCAM. According to the content, Qi-gong can also be categorized into the classes of Mind, Life, and Mind and Life (combined). For Mind Qi-gong, the practice has been focused on meditation and mind regulation; for Life Qi-gong, the focus is on the regulation of body functions; and for Mind and Life Qi-gong, practice is conducted with more of a balance of both mind and body function. Finally, Qi-gong routines were occupationally classified based on locations where they were developed (e.g., Shao Lin or Wu Dang, which is the name of a mountain) or function (e.g., self-healing, intelligence enhancement, and performance entertainment).

A few words should be said about the Qi-gong routine Master Chen performed. It is called Shao Lin-Nei Jin-Yi Chi-Chan, where "Shao Lin" is the name of one of the most famous temples in China, from which Wushu was developed; "Nei Jin" represents internal strength; "Yi Chi" means single finger; and "Chan" is a term in Buddhism meaning prolonged and intense contemplation. The routine represents one of the highest-level Qi-gongs in China. For many generations, it has been taught only within the temple to selected monks. Master Ah-Shui Que, a monk in the Southern Shao Lin Temple for many years, learned the secret. Because he was his family's only son, he eventually left the temple after 18 years of temple life and started a family in Shanghai, the largest city in China. He worked as sweeper for a long time and remained unknown for his exceptional Qi ability. When he miraculously cured many people with serious medical conditions, he became known in Shanghai and started to train some students. When the Chinese soccer team lost an important Asian game in the 1970s due mainly to physical disadvantages, the government looked for a training method, hoping to make soccer players become stronger quickly. After Master Que was recommended, he publicly displayed for the first time his incredible Qi ability (e.g., to guide a person without touching, to strike his head and body against a concrete wall). In 1978, he participated in the first-ever study to try to measure Qi. Master Que passed away in 1982. Many of his students have the ability to use external Qi for treatment and healing purposes. The routine itself became so popular in the late 1980s and early 1990s that millions of people in China practice it.

Mystery and Controversy Surrounding Qi-Gong

Because of the many unexplained phenomena associated with Qi and Qi-gong, the secrecy in the method of teaching, ties with religions, and a lack of measurability, Qi and Qi-gong have been full of mystery and controversy throughout their history. While efforts were made in the 1970s and the 1980s to measure Qi, and new evidence has accumulated in recent years (Lin, 1987; Lin & Chen, 2002; Liu et al., 1992), many practices and much research in Qi have been criticized as "pseudoscience" (Lin et al., 2000). Is Qi really a human super-ability that only a few people have, or is it an unknown ability that everyone has, or simply a trick? I believe that Qi, like muscular strength, is an ability and function that everyone has. With a little training, everyone can feel and benefit from Qi. As with Olympic weightlifting, in which only a few people, after many years of systematic training, become strong enough to participate, one needs many years of effective and demanding training to be able to reach the Qi level of Master Chen.

A quick experiment may help demonstrate the points. Sit down on a chair as shown in the position in figure 10.2*a*. Put your right arm in front of your body, bend your elbow 90°, and point your thumb toward your nose with four fingers together in a vertical position. Put your left palm under your right elbow, separated by about a 1-in. distance. Concentrate on the area between your elbow and the middle of your palm, where an acupuncture point called Lao Gong exists, for a few minutes and slightly move your elbow or palm. Do you feel something there? You may not feel anything or very much at this time, but try to remember what you felt even if you could not feel anything. Now, stand up and take the horse stance position illustrated in figure 10.2*b* and *c*.

a

b

c

FIGURE 10.2 Illustrations
for Qi experience.

1. Place your feet approximately shoulder-width apart.
2. Hold elbows and hands at the same level with your elbows by your side and your hands out in front of you.
3. Have palms facing an invisible point on the floor between the feet with wrists straight and fingers slightly apart, relaxed (your little, ring, middle and index fingers should be formed like a stair with the index finger at the highest point).
4. Bend your knees slightly with the knees above the toes, but not over them.
5. Turn toes slightly inward.
6. Keep upper body straight and head held up, but relaxed.

Keep this position for 5 to 10 min (the longer, the better; but maintaining the position can be difficult at first). Now return to sitting in the previous position in the chair and again concentrate on the area between your elbow and palm. You should feel something there now: a little warmth and something like a magnetic field. That is Qi! It is possible that you might still not feel anything, but just keep practicing the position of the horse stance for a week, for 15 min a day, and almost everyone can feel it.

It should be noted that, like other human abilities, for example running ability, the interindividual variability in Qi is large. Some can feel and even release Qi with minimal training, while others may require many years of practice to finally feel it. It is said that when reaching a very high level of Qi, one can easily feel the flow of Qi among meridians or channels in the body and the release of Qi through acupuncture points. Channels and acupuncture points are the foundation of traditional Chinese medicine and have been well supported by modern science. It is also said that when a person's Qi achieves that level, he or she can easily feel and identify health problems of others and treat their problems by releasing their Qi (Liu et al., 1992). As expected, effects and benefits of Qi have sometimes been exaggeration, either by those with a high-level Qi ability or by their followers. As a result, Qi masters were sometimes viewed as "gods," and practice of Qi-gong became a religious practice or a superstitious activity.

Qi-Gong and Health

While there is great interest in and need to understand Qi, perhaps the most important question to the general public is whether practicing Qi-gong can provide health benefits. The answer is definitely yes, as long as the practice follows appropriate guidelines. This is the same as with practicing any exercise-related training. A wealth of literature in English has shown the benefits of practicing Qi-gong. The following are just a few examples, and interested readers should view a complete resource list on the Qi-gong Institute Web site (www.qigonginstitute.org).

• Practicing Qi-gong has been shown to significantly improve the cardiovascular system, for example, to decrease systolic blood pressure and heart and respiration rates (Lee et al., 2000, 2002; Lee, Lee, & Kim, 2003; Mayer, 1999).

• Practicing Qi-gong can improve glucose metabolism through the benefits of the relaxation response and therefore may be a beneficial adjunctive treatment for individuals with type 2 diabetes (Iwao et al., 1999).

• Practicing Qi-gong has been shown to be beneficial in treating cancer, both in human subjects and in animal experiments (Chen & Yeung, 2002; Jones, 2001).

• Qi-gong, as a mind–body intervention, has been shown to have a beneficial role in the relaxation response and cognitive reconstructing in the treatment of headaches, insomnia, pain, and cardiovascular disorders (Tsang, Cheung, & Lak, 2002; Tsang et al., 2003; Wu et al., 1999).

• Practice of Qi-gong has been shown to have multifaceted health benefits (Liu et al., 1990; Sancier, 1996; Sancier & Holman, 2004).

It should be pointed out that a mountain of evidence is also to be found in the huge amount of Chinese literature. Although some of these studies were not tightly controlled, the health benefits of practicing Qi-gong are still strongly supported.

Meanwhile, since everything has two sides, it seems well to note some negatives. It has been reported in both historical and modern literature that the practice of Qi-gong, especially internal styles, may lead to "Zou Huo Ru Mo," which means "Qi-gong-induced mental disorders" (QIMD). According to a recent review (Ng, 1998), however, many so-called QIMD may be more appropriately labeled "Qi-gong-precipitated psychoses," where the practice of Qi-gong acts as a stressor or catalyst in vulnerable individuals. In practicing Qi-gong, as with participating in any exercise, one must observe scientific principles. Using long-distance running as an example, while people may significantly improve their cardiovascular function by running long distances, they may get injured or even die if they participate in a marathon race without systematic training or passing a medical qualification screening.

Qi-Gong and Aging

What about aging? Can the practice of Qi-gong lengthen one's life or delay aging? Most traditional Chinese medicine and wellness theories hold that one's maximal life span is fixed at birth. That is, one's parents determine one's maximal life span. Whether a person can reach his or her maximal life span, however, will be determined by that person's lifestyle, physical and psychological wellness, and adaptation to the environment. Life is like a candle with different lengths, many ancient Chinese wellness philosophers believe. If one lives an unhealthy lifestyle and cannot adapt to the environment, the candle will likely burn out first even if it is longer than others at the beginning. In contrast, if one lives a healthy lifestyle with a balanced mental status, the candle will likely last much longer even if it is a short one. Practicing Qi-gong regularly will bring peace to one's mind and help balance one's self in relation to the environment. As a result, practicing Qi-gong should help people reach their maximal life span. It should be noted that while receiving external Qi from a Qi master can help people overcome their health problems, those who often give of their Qi may be harming their own health. Although the idea is not based on scientific investigation, it is believed that some external Qi healers shortened their life spans by giving out too much Qi (Liu et al., 1992).

Qi Measurement and Challenges

To completely understand Qi, Qi-gong, and their healing effects on health and aging, we have to be able to understand the true nature of Qi. The ability to measure Qi is perhaps the most important step. As an old measurement saying has it, "Whatever exists at all exists in some amount. To know it thoroughly involves knowing its quantity as well as its quality." In fact, researchers both inside and outside of China have made efforts to measure Qi. It has been reported that an early attempt in the 1970s showed Qi to be detectable by temperature and infrared. Very recently, Sancier tried to monitor the effect of Qi-gong practice by using electrodermal measurements. Others (Litscher et al., 2001) have reported that Qi-gong practice appears able to objectify accompanying cerebral modulations. Overall, however, little is known about Qi. What is Qi? How is Qi developed? How is Qi released and received? Many of these critical questions remain unanswered. For more information about Qi measurement, interested readers are referred to an excellent review by Lin and Chen (2002).

There are major challenges in measuring Qi. First, there is a scarcity of convincing theory or hypotheses about the possible scientific foundation of Qi. Even Qi-gong masters have trouble explaining Qi and its true nature. Second, it is often difficult to identify true Qi healers, especially those who practice internal styles. Third, not too many known true Qi healers are willing to participate in scientific research because of potential harmful effects to their health. Finally, scientists in other disciplines have little knowledge about what (e.g., temperature, infrared rays, or something else) should be measured. As a result, most test equipment employed may not be able to measure Qi well, if at all.

Future Research Directions

To understand Qi and Qi-gong and to allow more people to benefit from this ancient yet powerful healing technique, systematic and multidisciplinary efforts are needed. Perhaps the first step is to confirm Qi-gong's healing effects using modern scientific research designs (e.g., random clinical trials) and valid and reliable outcome measures, and to publish the research findings in first-rate international research journals. Only in this way can mainstream Western medical researchers recognize the merit of Qi-gong. With confirmed scientific evidence, mass introduction of Qi-gong to the general public may become possible. With increased awareness of the benefits of Qi-gong, the number of practitioners and the research environment, including research funding and availability of Qi healers and participants, should significantly grow. Multidisciplinary expertise is definitely needed to measure and understand Qi, and specific equipment has to be developed for the purpose of measurement. Finally, the research focus should be on measuring external Qi, since the placebo effect, which is often associated with Qi studies, can be relatively easy to block.

Part III

New Measurement Methods and Techniques

Chapter 11

Common Shape Models for Trend Curves

Roderick P. McDonald

Suppose we have a single response variable $V(t)$ measured at r times on N subjects. The observations $V(t_1), \ldots, V(t_r)$ for any individual i may represent growth, learning, decline of a function with age, or any measure of interest that may show change with time. Any systematic change over time may be modeled by a smooth function, with departures from the curve regarded as random errors. A smooth function fitted to such an individual set of observations is a *trend* curve. We might in addition have some other measures, $X_1, \ldots, X_p$, that either characterize the individual subjects or are measures of controlled experimental conditions or uncontrolled environmental conditions. We will refer to the latter as contextual variables. Given a set of N curves fitted to the subjects, we can treat their parameters as characterizing (individual differences in) the subjects and do correlational work or causal analysis—analysis of variance or structural equation modeling—relating these parameters to the contextual variables.

To make these abstractions more concrete, consider the data given in table 11.1 and graphed in figure 11.1.

This data set represents responses from 18 subjects over 36 trials on a skill-learning test (Yates & McDonald, 1966). Perhaps the most striking feature of the data is the fact that, in an intuitively compelling sense, the underlying trends appear to have a common shape, or, we might say, to follow a common law. At the same time, there are visible individual differences in the beginning and in the final levels of performance. These differences are possibly large enough to preclude fitting a common curve to all 18 sets of responses.

Classical regression methods can easily be applied to fit either an individual trend curve or, simultaneously, a set of N trend curves, if we are content to represent the curve by a simple polynomial. Applying a simple regression program to the data in table 11.1 establishes that each of the trends is well approximated by a quadratic curve

Table 11.1 McDonald-Yates Learning Data

SUBJECT

Trial	1	2	3	4	5	6	7	8	9	10	11	12	13	14	15	16	17	18
1	9	16	10	14	7	7	10	11	7	7	8	12	12	5	11	8	10	11
2	16	19	12	19	12	7	13	13	11	12	14	19	11	12	8	18	11	14
3	16	17	14	14	15	7	15	13	7	12	6	14	15	13	12	7	12	17
4	14	11	14	18	15	11	19	11	12	12	12	19	20	15	11	11	12	18
5	17	21	20	18	17	11	22	15	12	15	17	20	15	8	13	19	18	19
6	18	21	21	18	19	11	22	15	7	15	14	23	14	14	14	14	19	21
7	18	19	21	19	16	16	23	16	12	3	15	26	15	13	12	15	20	21
8	18	19	23	23	20	16	25	16	15	10	21	23	15	20	17	17	20	22
9	23	21	23	21	15	18	26	17	13	12	20	29	17	23	15	20	21	23
10	24	19	24	23	21	18	26	19	14	13	14	29	20	17	17	20	23	23
11	25	23	19	25	19	21	27	13	17	13	21	31	20	18	17	23	22	20
12	26	23	25	21	17	21	25	17	15	14	21	28	26	17	17	25	23	23
13	26	23	27	26	17	25	27	13	16	14	21	31	25	20	20	25	23	26
14	26	23	28	25	21	25	28	19	17	14	21	33	27	20	21	25	24	29
15	29	25	30	26	21	25	29	19	17	15	21	31	28	21	21	25	26	27
16	31	25	28	27	21	25	30	17	20	17	22	35	30	24	21	25	28	28

17	29	29	27	19	25	27	35	22	16	21	19	31	29	22	28	30	28	26
18	29	28	26	21	25	31	34	22	17	17	19	35	27	19	28	31	28	27
19	30	29	28	21	27	31	38	22	17	18	19	33	32	15	28	32	28	30
20	31	30	28	21	29	31	34	23	13	20	21	31	31	22	28	34	29	33
21	29	31	28	21	23	29	39	23	13	19	23	33	30	23	28	30	32	34
22	29	31	28	21	29	29	40	23	19	19	21	35	32	23	30	35	33	34
23	31	32	30	21	28	22	41	27	15	19	21	34	27	23	30	35	33	30
24	27	32	31	23	29	26	43	24	17	20	22	35	33	23	25	36	33	34
25	31	31	31	23	30	32	34	26	17	29	14	35	34	25	31	37	33	36
26	31	31	30	24	25	32	43	26	17	23	23	38	32	25	32	39	34	36
27	31	33	31	24	31	32	41	24	19	23	15	39	32	25	35	35	33	32
28	32	32	31	21	31	31	41	26	19	21	22	39	30	25	31	39	34	37
29	32	36	33	25	31	32	44	28	13	21	23	40	32	26	31	40	34	38
30	30	35	33	26	31	40	44	27	15	16	23	41	34	26	33	41	35	35
31	32	35	31	25	32	32	44	28	19	21	23	36	36	29	30	39	35	38
32	32	34	34	26	30	32	42	28	19	23	23	35	34	29	29	35	36	38
33	34	36	33	28	33	29	45	28	20	24	23	41	38	30	34	41	36	36
34	34	35	35	27	35	30	41	28	19	18	23	42	38	26	35	41	36	38
35	36	34	34	28	33	30	44	28	20	21	23	44	39	27	34	42	36	38
36	35	36	35	23	31	33	48	28	19	25	23	44	40	27	31	41	38	39

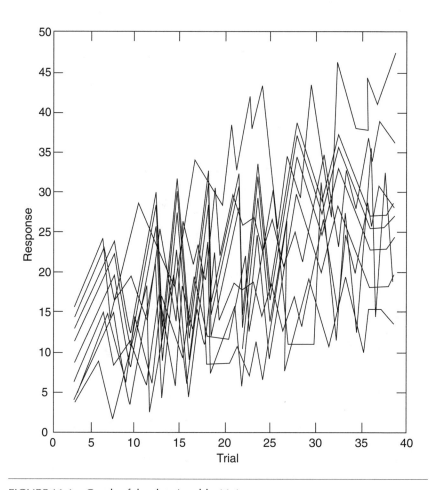

FIGURE 11.1   Graph of the data in table 11.1.

$$V_i(t) = \beta_{0i} + \beta_{1i}t + \beta_{2i}t^2 + E_{it} \qquad (11.1)$$

with regression parameters β_{0i}, β_{1i}, β_{2i}, and a regression residual E_{it}. In the context of learning, the use of a polynomial is open to a strong theoretical objection, namely that the quadratic curve may give increasing values over a limited range of times/trials but these values must eventually start to decrease instead of providing an asymptote to the learning process. In other contexts—aging is an obvious one—there may be no theoretical basis for a choice of curve that is preferable to the simple polynomial.

Rather more complex regression methods are necessary to fit a theoretically appropriate curve to the data in table 11.1. As a suitable curve for a learning process we might choose the exponential

$$V_i(t) = \alpha_i - (\alpha_i - \beta_i)\exp(-\gamma_i(t-1)) \qquad (11.2)$$

The form of this function is chosen, following Browne (1993), so that each parameter has a simple interpretation, with α the asymptote, β the baseline response, and γ the curvature. McDonald (2004) shows how to adapt a structural equation modeling program to fit strictly nonlinear trend curves such as the exponential. Using this method we find that the exponential curve and the quadratic curve are virtually indistinguishable for the sample data set. See columns E and Q in table 11.2.

TABLE 11.2   Fitted Curves for Subject 1

Tr	V	E	Q	Qa	Qs	M	Q_M	E_M
1	9	11.1	11.7	11.8	13.1	9.7	11.3	10.7
2	16	12.7	13.1	13.1	14.3	13.4	12.4	12.0
3	16	14.2	14.4	14.4	15.6	12.6	13.4	13.2
4	14	15.7	15.7	15.7	16.7	14.2	13.5	14.4
5	17	17.0	16.9	17.0	17.9	16.5	15.5	15.5
6	18	18.3	18.2	18.2	19.0	16.7	16.4	16.5
7	18	19.6	19.3	19.3	20.1	16.7	17.4	17.5
8	18	20.8	20.5	20.5	21.2	18.9	18.2	18.5
9	23	21.9	21.6	21.6	22.2	19.8	19.2	19.4
10	24	23.0	22.7	22.7	23.2	20.2	20.0	20.3
11	25	24.0	23.7	23.7	24.2	20.8	20.9	21.1
12	26	25.0	24.7	24.7	25.1	21.3	21.7	21.9
13	26	25.9	25.7	25.7	26.0	22.5	22.4	22.6
14	26	26.8	26.6	26.5	26.9	23.7	23.2	23.4
15	29	27.7	27.5	27.5	27.7	24.2	23.9	24.0
16	31	28.5	28.4	28.3	28.5	25.2	24.6	24.7
17	26	29.2	29.2	29.2	29.3	25.7	25.3	25.3
18	27	30.0	30.0	30.0	30.0	25.8	25.9	25.9
19	30	30.7	30.8	30.7	30.7	26.6	26.5	26.5
20	33	31.4	31.5	31.4	31.4	27.2	27.1	27.1
21	34	32.0	32.2	32.1	32.0	27.1	27.6	27.5
22	34	32.6	32.8	32.8	32.6	28.4	28.1	28.0
23	30	33.2	33.4	33.4	33.2	27.7	28.6	28.5
24	34	33.7	34.0	34.0	33.7	28.5	29.1	28.9
25	36	34.3	34.6	34.5	34.3	29.4	29.5	29.3
26	36	34.8	35.1	35.0	34.7	30.1	30.0	29.7

(continued)

TABLE 11.2 *(continued)*

Tr	V	E	Q	Qa	Qs	M	Q_M	E_M
27	32	35.3	35.5	35.5	35.2	29.7	30.3	30.1
28	37	35.7	36.0	36.0	35.6	30.2	30.7	30.5
29	38	36.1	36.4	36.4	36.0	31.0	31.0	30.8
30	35	36.6	36.7	36.7	36.3	30.9	31.3	31.2
31	38	37.0	37.0	37.1	36.6	31.8	31.6	31.5
32	38	37.3	37.3	37.4	36.9	31.1	31.8	31.8
33	36	37.7	37.6	37.6	37.2	32.9	32.1	32.1
34	38	38.0	37.8	37.9	37.4	32.2	32.2	32.6
35	38	38.4	38.0	38.1	37.6	32.8	32.4	32.6
36	39	38.7	38.1	38.2	37.7	33.1	32.5	32.9

V = Subject 1—score on skill test; E = S1—exponential curve; Q = S1—quadratic curve; Qa = S1—scale and origin transformation of quadratic reference curve; Qs = S1—scale transformation of quadratic reference curve; M = means of scores over subjects; Q_M = quadratic (reference) curve fitted to means; E_M = exponential (reference) curve fitted to means.

It is easy to show—as we would perhaps expect by intuition—that if we wish to fit a common curve to all N sets of data, we may fit the curve by regression methods just using the mean response of the N subjects at each of the r times. Carrying out this procedure in our example gives the means in the column headed M in table 11.2 and, also, the corresponding points Q_M of the fitted quadratic curve and the points E_M of the fitted exponential curve. While for some scientific purposes we would be very pleased if a common curve sufficiently fit all the individuals in a data set, we would not be surprised to find in an application that it did not. In the present example the mean squared errors of the individuals about the common curve—column MQ of table 11.3—are much larger than their mean squared errors about their own fitted curves (column IQ of table 11.3). These are the means over the 36 trials of the squared differences between the individual responses and, respectively, the points of the quadratic fitted to the means (Q_M in table 11.2) and the points of the individually fitted quadratics. The rest of the information in table 11.3 concerns the common shape models, to which we now turn.

Common Shape Models

A number of approaches to the problem of individual differences in trend curves can be found in the literature (Browne, 1993; Browne & du Toit, 1991; McArdle & Bell, 2002; McDonald, 1965 and in press; Meredith & Tisak, 1990; Tucker, 1958). The reader is referred to McDonald (2004) for a review of some of these, including a more technical account of the common shape models to be discussed here.

TABLE 11.3   Mean Squared Errors for Reference Curve Models

S	IQ	SOQ	SQ	IE	SOE	SE	MQ
1	3.1	3.1	3.4	3.1	3.4	3.7	19.9
2	3.1	3.8	3.9	3.1	4.5	4.6	11.7
3	5.4	3.4	4.6	3.3	3.9	5.1	35.3
4	3.3	3.3	5.1	3.3	3.5	5.3	6.9
5	4.7	5.0	5.9	4.7	5.0	5.8	20.7
6	3.6	3.6	10.9	3.3	3.3	10.8	13.9
7	4.1	4.2	4.2	3.7	4.0	4.0	44.5
8	4.6	4.7	6.9	4.6	4.8	7.1	52.7
9	5.7	6.0	6.0	5.8	6.2	6.3	61.7
10	5.7	5.8	7.2	5.7	5.8	7.1	113.0
11	3.9	3.9	4.0	3.6	3.8	3.9	14.3
12	4.9	4.9	5.0	4.8	5.3	5.3	95.6
13	8.8	9.4	9.4	9.2	9.8	9.8	10.4
14	5.4	5.4	6.7	5.4	5.0	6.4	7.5
15	2.2	2.2	2.2	2.1	2.2	2.2	27.8
16	4.2	4.2	4.6	4.0	4.2	4.6	5.4
17	1.4	1.5	1.8	1.3	1.3	1.6	6.7
18	2.6	2.7	4.0	2.2	2.4	3.7	7.8

IQ = individual fit, quadratic curve; SOQ = scale and origin transformation, quadratic; SQ = scale transformation, quadratic; IE = individual fit, exponential curve; SOE = scale and origin transformation, exponential; SE = scale transformation, exponential; MQ = average errors about quadratic (reference) model fitted to means.

One advantage of the polynomial model—in our example, the quadratic model (equation 11.1)—over strictly nonlinear models such as the exponential model (equation 11.2), is that the polynomial is a weighted sum of components. It is this property that allows the application of classical regression models to fit the individual curves. It is also true that the components in the polynomial, $1, t, t^2, \ldots$, are independent of the data, $V_i(t)$. The remainder of this discussion concerns the possibility of describing a set of individual trend curves as transformations of a reference curve derived from the data. Like the polynomial, the models to be considered have the advantage of being linear and additive in the parameters describing the individuals.

Consider first the simple linear additive model:

$$V_i(t_j) = \beta_{0i} + \beta_{1i}t_j + E_{ij} \tag{11.3}$$

This model has three parameters, easily estimated by regression methods. These are the regression constant, representing baseline performance, β_{0i}; the slope of the straight line trend, representing gain per trial, β_{1i}; and error variance. These might correspond to attributes such as, respectively, prior knowledge, cognitive ability, and carelessness. With just this simple model, given estimates of these parameters we can investigate their relationship with contextual variables. In some applications (e.g., learning trials as in our example), the time line has a natural origin. In other applications it may be reasonable to regard the origin of the time measure as arbitrary.

Suppose we accept the origin of time as given. We then might wish to check how closely β_{0i} and β_{1i} are proportional in the simple linear model (equation 11.3). Their agreement can be measured by Burt's (1948) classical *congruence coefficient*, a normalized raw product moment

$$c = [\Sigma\beta_0\beta_1]/[(\Sigma\beta_0^2)(\Sigma\beta_1^2)^{1/2} \tag{11.4}$$

or perhaps by the corresponding expression for their deviations from their means,

$$cs = [\Sigma\delta_0\delta_1]/[(\Sigma\delta_0^2)(\Sigma\delta_1^2)^{1/2} \tag{11.5}$$

In the case where the origin of time is arbitrary, the relation between δ_{0i} and δ_{1i} depends on the origin chosen. We consider choosing a new origin t_0 for time, so that on the new scale, denoted by $t^* = t - t_0$,

$$V_i(t_j) = (\beta_{0i} + \beta_{1i}t_0) + \beta_{1i}t^* + E_{ij} \tag{11.6}$$

and, writing V_j for the mean at time t_j,

$$V_i(t_j) - V_j = (\delta_{0i} + \delta_{1i}t_0) + \delta_{1i}t_j^* + Eij = \delta_{0i}^* + \delta_{1i}t_j^* + E_{ij} \tag{11.7}$$

It may be shown that

$$\Sigma\delta_{0i}^{*2} + \Sigma\{\delta_{0i} + \delta_{1i}t\}^2 \tag{11.8}$$

is a minimum at $t^* = t_0$, where

$$t_0 = -[\Sigma\delta_{0i}\delta_{1i}]/(\Sigma\delta_{1i}^2) \tag{11.9}$$

and with that choice of origin

$$\Sigma\delta_{0i}^*\delta_{1i} = 0 \tag{11.10}$$

Also, at the new origin,

$$\Sigma\delta_{0i}^{*2} = \Sigma\delta_{0i}^2 - [\Sigma\delta_{0i}\delta_{1i}]^2/[\Sigma\delta_{1i}^2] \tag{11.11}$$

This last quantity is zero if and only if $\delta_{0i} = \alpha\delta_{1i}$, that is, if the coefficients are proportional. We might reasonably refer to the point t_0 as the focal time—the time point at which the lines are as close together as possible, resembling the focusing of light rays. In an ideal case the lines may have a unique intersection at a point on the time axis within the time interval of the model. It is more likely that the lines will intersect imperfectly. They may intersect inside or outside the time interval in which the data lie. They might in some cases be parallel, to a close approximation. It will generally be of interest to study the location of the point of intersection.

Depending on the empirical context, in some applications we may be content with the estimates of δ_{0i} and δ_{1i} (and their means and variances), together with the individual error variances, regarded as baseline, rate, and error measures characterizing the subjects. In others we may prefer to rescale time and use linearly independent "uncorrelated" parameters δ_{0i}^* and δ_{1i}^* to characterize the subjects, or at least to estimate the time point at which the lines come to an approximate focus and study its relationship with the time range of the study. Clearly the covariances of these parameters are not invariants of the model, a fact that may be of importance in some applications involving the relationships between these and other variables. This fact does not seem to have been widely recognized.

The results just given do not generalize readily to nonlinear trend curves. However, the recognition that the coefficients, δ_{0i}, δ_{1i}, may be linearly dependent does suggest a further possibility, as follows:

We suppose that each individual trend curve can be written as a transformation, in the form of a simple translation and dilation, of a single reference curve $[M(t_1), \ldots , M(t_r)]$. This might be the set of r fitted values Q_M, or E_M, the quadratic and exponential functions fitted to the means and listed in table 11.2.

That is, we write

$$V_i(t) = \gamma_{0i} + \gamma_{1i}M(t) + E_{it} \qquad (11.12)$$

(And conversely there is a transformation of each individual's response curve into the reference curve.) We might also attempt to account for the data by a scale change only, with

$$\underline{V} = \gamma_{1i}M(t) + E_{it} \qquad (11.13)$$

In one meaning of the term "shape," all the curves have the same shape, differing only by a change of unit and possibly origin—and there is a clear sense in which they follow the same law. If this model fits, it suffices to relate the scale parameters γ_{0i}, γ_{1i} to other measures on the individuals (together with the individual error variance, if desired).

A simple two-stage procedure for this model (McDonald, 1967) consists in fitting a curve to the means over subjects at each trial to give the common model, and using this as a reference curve in equation 11.12 or equation 11.13.

The method applies to a simple additive model such as the polynomial or to a strictly nonlinear model, fitting a prescribed curve such as the exponential to the means as a common model and using this as a reference curve. Whereas in the simple additive model the resulting individual curves follow the same function as the reference curve, this is not generally true of any nonlinear function we might choose. Browne (1993) makes use of the property that a number of commonly used nonlinear curves are scale free in the sense that multiplying the function by a constant is equivalent to multiplying the parameters by that constant. Fortunately the exponential function satisfies the even more general condition that

$$c + k[\alpha - (\alpha - \beta)\exp(-\gamma(t-1))] = \alpha^* - (\alpha^* - \beta^*)\exp[-\gamma(t-1)] \qquad (11.14)$$

where

$$\alpha^* = c + k\alpha \qquad (11.15)$$

and

$$\beta^* = c + k\beta \qquad (11.16)$$

That is, rescaling the exponential function gives an exponential function, so we unhesitatingly use it as a common model and reference curve to obtain individual exponential curves by equation 11.12 or equation 11.13.

The reference curve procedure has been applied using (a) the change of origin and scale and (b) the change of scale only, to (i) the quadratic common model and (ii) the exponential common model previously fitted. Table 11.3, which already contains the mean squared errors for the individuals about their own curves and about the common curve, gives the mean squared error of fit to the individuals for these four models. The results are remarkably good. From such data we are also able to detect any individuals whose curves do not appear to follow the common law (none in the example), and those whose data require a change of origin as well as a change of scale (subjects 6 and 8).

Conclusions

It has been shown that a simple way to obtain a small number of parameters describing individual differences in a set of trend curves is to use a common model as a basic, data-dependent reference curve and to derive individual curves from it by a change of scale and possibly a change of origin. In the example given here, the data follow a common law of learning. Browne (1993) gives a contrasting case of learning data in which subjects clearly follow different laws, some appearing to have their learning complicated by an inhibitory process in the late trials while others do not. It is an important question of empirical fact, testable by the common shape model, whether individual trend curves follow a common law. If they do, this fact simplifies further work relating them to contextual variables.

Chapter 12

Emergent Technologies and Remote Clinical Assessment

Leigh W. Jerome, PhD

Rapid developments in science and technology are profoundly affecting diagnostics, assessment and treatment in health care, and health promotion. Continued innovation, over the next several decades, will precipitate shifts in our basic understanding of the causes and course of illnesses and thereby influence human perspective on relationships, self-concept, and ethics. Keeping pace with rapid innovation means identifying fledgling changes toward anticipating new possibilities and preparing for related societal implications.

The goal is to envision and create the future, as we know it should be, by constructing opportunities that foster prevention, healthier lifestyles, and sound ethical processes. Technology will not cure all of society's ills. In fact, technological advances will require vigilance and directed efforts to ensure that they do not create as many problems as they resolve. The challenge is to develop innovations as well as preparedness to respond to their impacts, to develop an ecosystem in which technology-supported progress co-evolves in response to knowledge about one another's needs and constraints.

Recognizing the depth and breadth of coming technological changes will allow us to create a vision for our preferred future and will prepare us to better guide our patients, teach our students, and expand the landscape for productive aging. Understanding where changes are developing allows us to anticipate where *we* want to be.

One important concept related to assessment and emergent technologies is convergence. The big convergence is the blending of traditional media (print, television, radio, and film) with the "new technologies"—namely, cable television, the Internet, and data-casting. We also see many convergences of various *specific* technologies. Music and videos can be downloaded directly from the computer. Photos can be taken and transmitted as part of your cell phone technology. You can go online to shop, read the newspaper, or talk to your physician.

We are witnessing an enormous convergence of health care, science, and telecommunications. This convergence is reflected in the diversity of technologies emerging for assessment and intervention with geriatric populations. This chapter deals with technologies that are being developed and implemented to promote the physical and emotional well-being of older adult populations.

It is estimated that by 2030, people ages 65 and over will represent 22% of the population in this country (Czaja, 2003). Many of these individuals will be over the age of 75 years, and many will need care. At the same time that the U.S. population is aging, technology is becoming integral to everyday life. In a nation of 275 million people, about 68% of households own a PC, and approximately 175 million Americans use the Internet.

About 100 million Americans use the Internet to learn about their health care issues and search for facts about treatments, medications, and alternative approaches to traditional care. With the communications infrastructure now in place to support technology and the increasing affordability of technology solutions, options that have previously been cost prohibitive are becoming cost-effective. As patients become more accustomed to electronic communication, they are beginning to want more than simple access to general information. They are starting to demand electronic health options.

Patient education via the Internet allows patients to learn about their disease. A 2002 Pew study showed that 31% of all Americans said they would turn first to the Internet for information. The Internet provides a forum for *patients,* but also a place where *caregivers* can go for information and support. Online opportunities can provide greater understanding of chronic disease, disability, and aging and thus promote emotional well-being and physical health.

Convergences are allowing us to overcome geographical boundaries through hub and spokes networking. In the traditional hub and spokes model, the hub is a central organization such as an academic or medical setting. The hub is connected, via telecommunications, to numerous spokes—for instance, remote health clinics—that are distributed across broad geographical regions. They may be connected through store-and-forward (e-mail), audio connectivity, interactive televideo connections, or more than one of these.

Convergence blurs the boundaries of traditional assessment *environments.* Historically, assessments have been divided into different specialty areas—educational assessment in the schools, clinical assessment in clinical environments, and personnel assessment at the workplace. Assessments were conducted as discrete testing events without a lot of overlap between the specialty areas.

The importance of assessments becoming more continuous, outcomes based, authentic, and multifaceted is recognized across the specialty areas. Educational, health care, and workplace environments are being connected to provide seamless avenues for assessment, learning, and the provision of clinical services.

We will look first at how workplace assessment and education are being integrated. The knowledge required to maintain a job in many occupations is changing so fast that an estimated 50% of all employees' skills may become outdated within three to five years (Bennett, 2001). Meanwhile, longevity and later retirement ages are making education a lifelong process.

Suppose that a change in the way a company does business is going to require greater computer literacy. Assessment and training can occur within the employment environment, directly at the point of need. Online evaluations can assess a person's technology literacy as well as any anxiety or physical challenges that might affect the new learning. Needed skill sets are thus identified.

The Internet provides enormous potential for the seamless embedding of assessment in instruction. Online modules can be customized to address particular skills as needed for individuals. An electronic portfolio can then capture growth over time—containing audio, video, graphics, and text. Electronic portfolios are able to use hypertext links between documents to reflect multiple perspectives. When instructional exercises are responded to electronically, the responses can be recorded, leaving a continuous learning trace.

Expert systems are being developed to organize the individual's information—to determine what instruction would be best to present next and to document what competencies the individual has achieved. So, assessment is tooled to provide

- knowledge facilitation, helping people develop their skills; and
- knowledge certification, to document that they have developed the skills.

Going a step further, health care is converging with on-the-job assessment and education. Assessment is becoming a dynamic process rather than a singular event, through phases of life and for a variety of needs. Neuropsychological and IQ testing have traditionally been considered clinical assessments, while performance has been an academic measure. But there are *many* kinds of knowledge that help people respond to their environment effectively, including visual-spatial, bodily-kinesthetic, musical-rhythmical, and interpersonal skills. Further, assessments of attitudes, motivation, time management, anxiety, concentration, information processing, and test strategies are converging with traditional cognitive assessments.

Testing is broadening to include the full range of relevant domains—with assessment occurring at the point of need. Electronic assessment data from multiple sources can provide clear documentation of baseline functioning so that earlier detection of and intervention into disease processes can occur. An example may help illustrate the value of these converging events. *Alzheimer's disease* affects more than 4 million Americans. A new high-resolution functional magnetic resonance imaging (MRI) technique has been developed that can pinpoint changes in brain activity that may underlie memory impairment, even before structural damage occurs (Norfray & Provenzale, 2004). The use of the MRI to precisely map blood oxygenation in the brain at rest has implications for differential diagnosis and memory loss treatment in the early stages of

Alzheimer's. Early detection can provide an opportunity to establish a baseline of cognitive functioning, quality of life, occupational risk areas, and emotional resources, as well as providing ongoing opportunities for the development of appropriate remediation treatments. Ongoing measurement can then identify subsequent cognitive slippages and medical events to provide more effective treatment and remediation opportunities. The value of new techniques such as the functional MRI is leveraged when it is paired with traditional dementia assessment and when the full record is electronic, is dynamic, and can be accessed and augmented online.

The five domain areas of a comprehensive geriatric assessment (CGA) are

- physical health,
- mental health and emotional functioning,
- social and economic status,
- functional status, and
- environmental characteristics.

The quantification of these domains, especially physical health, is difficult, but critical to geriatric assessment. Comprehensive geriatric assessment is being improved through emerging technological developments.

There are two emerging technologies that are overarching in their influence: wireless and broadband. It is predicted that within a few years, more people will be accessing the Internet through wireless devices than through the use of wired computers at home or the office. Wireless offers portability and convenience. Mobile and wireless technologies are enabling health care professionals to obtain knowledge immediately, whenever and wherever they need it.

Broadband is the high-speed transmission of digital data. With broadband, there is no need for dial-up because the connection is constant. Further, the speed at which one can send and receive information is greatly increased. Finally, the quality and quantity of data accessible to those with broadband far exceed what is available without it. Broadband makes the real-time transmission of multimedia data possible, without distortion.

The digitizing of images opens new ways of storing, retrieving, and distributing images. Digital images are inexpensive, and they maintain consistent image quality regardless of duplications. There is no processing time required, so the digital images are available immediately. The immediacy of digital images has significant value for emergency and remote care situations. Digital images are easily transmitted for consultative and diagnostic evaluations.

For example, state-of-the-art cardiac imaging equipment gives us the ability to better view and treat coronary artery blockages that are antecedents to heart attacks or other serious heart system damage (Hong et al., 2001). New computer tomography can produce incredibly sharp 3-D cardiac images that allow physicians to spot plaque and dangerous narrowings in the coronary vessels. Previously, examinations of the heart were generally conducted through the invasive insertion of a catheter. New imaging techniques promise to help the cardiovascular team with the prevention, diagnosis, treatment, and rehabilitation of cardiovascular patients.

The expansion of computing and telecommunications has been accompanied by greatly enhanced databases, widespread connectivity, an enormous amount of online data, and hypertext. Consider the electronic medical record. It has historically been a collection place for clinical notes, consults, and results reflecting things like lab, radiology, and pharmacology information. The promise is that digitizing medical records will change the medical record. Rather than a static, paper-based inventory it will become a dynamic, ever-evolving document, rich with imported data including digital images, video clips, and vital signs downloaded directly from biosensors. Clinicians will have access to clinical history; the real-time transfer of labs and pharmacological data; clinical images; video clips; and results of specialists' consults—and these records will be available anywhere and anytime. Hypertext will provide direct links to clinical practice guidelines and emerging techniques, clinical trials, and research. It will generate reminders, alerts, and patient care plans—as well as take-home information. Integrated information systems will allow access to lifelong, portable, comprehensive health records for each person.

The value can be seen, for example, in a project that has been initiated to provide a testament to the baseline functioning of elderly individuals in decline. The project's goal is to provide a document of a person's views about his or her own course of treatment and to promote basic human dignity in the individual's long-term care. Live-action, multimedia videos with semistructured interviews are being developed (Allen, 2003) to be integrated into an electronic record. The video clips can be initiated at the earliest stages of dementia detection and can accompany the individual throughout the care system and be updated as desired. A clear baseline and reliable monitoring assist in the early detection of problems. Prompt detection is essential to early intervention and the prevention of more complicated case presentations. A critical diagnostic factor for disease in the elderly is the deterioration of functional independence. This is a reliable marker of unrecognized disease or inadequate management (Kitwood, 1997).

A biosensor is an analytical device that converts a biological response into an electrical signal. There are many emerging applications including home health care, emergency alerting, and biofeedback. It is estimated that between 8% and 10% of the adult population over 65 years old require some kind of *home care*. This number rises to 36% for those over 80. Home monitoring systems are becoming two-way triage systems, linking patients to primary care facilities through home monitoring stations. This is especially useful technology for rehabilitation, elderly patients, and individuals with chronic conditions such as asthma, diabetes, and heart disease. Vital signs, blood glucose, weight, oxygen saturation, and electrocardiogram information can be monitored automatically, depending on patient need. Medication dispensing can be automated, with alerting as to when it is due. Patients become more involved in their own treatment by inputting data and accessing feedback about status and progress. Biosensors can give timely warnings of illness so people can seek help early, when intervention will do the most good.

Medical devices and innovative technologies are converging toward a prevention-oriented, consumer-driven model including customized wearable devices, electronic patient records, and wireless Internet-linked systems. The goal is to provide convenient and user-friendly health care at the point of need. Assessment is changing from episodic intervention in acute conditions toward an integrated health care approach, with a focus on prevention and early intervention.

Personal monitoring systems are also being developed that can give people personalized advice, feedback, and reinforcement. This will provide greater opportunity for disease prevention strategies and the promotion of healthy lifestyle development.

In the future, handheld biosensors, using a wide range of advanced technologies, will make it convenient for people to track a variety of vital information. Athletes will be able to get an edge with wearable biosensors that monitor heart rate, respiration, hydration, and temperature through a shirt or some other nonintrusive product. One can easily imagine future generations of body and equipment sensors that will precisely record all details of a player's movements, for example in tennis. These details could then be downloaded into systems that will analyze the motions, compare them with motions in previous games, and produce high-quality graphical displays to help the player's game.

Biosensors have other potential uses. Imagine a virtual environment that would react to input from biosensors. This is being researched with input devices such as gloves and body suits. Eye movement also has important possibilities, such that convergence of the eyes on a specific object in a virtual environment will select that object. Devices that offer this kind of assistance show promise for individuals with spinal cord injuries or other nervous system disorders. These devices will allow people with disabilities to operate machines and perform routine tasks with hands-free instrumentation, controlled either by small muscle movements, such as a blink of an eye (electromyography), or by brain activity (electroencephalography).

Computer games provide a medium that engages people for long periods of time. People usually return to the same game many times over. There are obvious lessons here for the developers of digitally based diagnostic and training materials. Gaming is already being developed for individuals with brain injuries, psychosis, eating disorders, and phobias and for behavioral health and prevention applications. Eventually, the convergence of biosensors and behavioral health will help us evolve games that synthesize entertainment with diagnostics, learning, and mastery to gently motivate improved health.

This leads to the construct of captology, the study of where computing technology and persuasion overlap. This term, coined by the Stanford Persuasive

Technology Lab, refers to the study of computers as persuasive technologies, including the design, theory, and analysis of interactive and computing products created for the purpose of changing people's attitudes or behaviors. At Stanford, investigations of these persuasive technologies are exploring how to motivate people to make positive behavior changes. For example, a computer doll simulates how hard it is to care for a baby. It cries obnoxiously at irregular intervals and also registers acts of abuse. The doll is designed to persuade teenagers to be sexually responsible. Another example is a product that tracks grocery purchases via bar-code scanners and club cards and then mails out coupons to correct people's diet deficiencies.

Practitioners are increasingly using the Internet and audio and video conferencing solutions to augment clinical services. Electronic technologies are being used to take histories, conduct consultations, and monitor recovery and rehabilitation progress as well as to provide patient education. Sixty percent of individuals over the age of 75 report feelings of loneliness and social isolation (Webb, 2003). Communication technologies can benefit the elderly and chronically ill by reducing isolation and loneliness and improving security. E-mail, cell phones, electronic shopping services, and online social opportunities can serve as important tools in reducing social isolation by enhancing contact with family, friends, and health care professionals (Czaja, 2003). Options that bring seniors back into the community promote health. Video conferencing for people who have recently undergone surgery combines the benefit of face-to-face communications with the convenience of a home visit. Assessments for quality of life, pain, activities of daily living, depression, anxiety, and pain management have been successfully conducted utilizing interactive televideo conferencing. Remote physiotherapy has been effectively performed (Baker, 2000). Streaming video (one-way transfer of video) will eventually be integrated as a desktop solution to enhance consultation and education instruction and to convey ultrasound, echocardiograms, and videos of motion such as tremors or gait abnormalities. Vocational counseling via real-time video conferencing can enable support to be provided in a timely and convenient manner for effective home rehabilitation. Digital imaging can provide remote surveillance of skin health.

Caregiver variables affect all aspects of the older individual's life. Social support for caregivers can be improved through online support groups and assistance, when needed, via interactive video conferencing (DeLeo, Carollo, & Buono, 1995). High-bandwidth communication technologies can also help the elderly to continue to engage in meaningful activity, and encourage family and community participation even when mobility problems interfere with travel. For example, an older woman living alone at some distance from her child and grandchild may be able to do virtual after-school child care, providing basic monitoring and activities such as reading or helping out with homework.

Until recently the only widely used bionic device was a pacemaker. However, many research groups have recently reinvested themselves in bionics research.

For example, a neural prosthetic very similar to a pacemaker has recently gained FDA approval. Instead of preventing heart attacks, it prevents seizures (Fanselow et al., 2000). Over 300,000 Americans have age-related macular degeneration; 28 million Americans have hearing loss (Johnson & Danhauer, 2002). The convergence of microelectronics and medicine is offering new hope through computer chip prostheses.

Research is exploring the connecting of a charge-coupled device (CCD) to the human brain (Shur et al., 2002). A CCD is what is generally used in the more expensive camcorders or in telescopes. The human eye is able to capture only roughly 30% of the available light. Charge-coupled devices, on the other hand, are able to capture almost 99.9% of the available light. If these devices are successful, people with failing vision may be able to recapture their sight.

Hearing aids have been in use for many years now in one form or another; but now, rather than just magnifying the sound, devices are being developed to replace the damaged part of the ear. Cochlear implant can already help about 10% of formerly deaf persons to hear and understand voices well enough to hold conversations (Kurzweil, 1999). Many replacements are somewhat limited at this point, but they are the beginning of a new world of bionics.

Theodore Berger (2001) at the University of Southern California has begun trials with the world's first brain prosthesis—an implantable memory. The hippocampus is involved in the formation of long-term memory. While it is not known how the hippocampus encodes information, it is known that electrical current stimulation corresponds to a set of outputs. On the basis of this information, Berger and his group have built a mathematical model and programmed the model on a chip. The chip will interface with the brain through two sets of implantable electrodes. One set of electrodes detects the activity coming in from the rest of the brain, and the other sends instructions back out to the brain. In short, the prosthesis is being developed to reestablish the ability to store new memories.

Virtual reality is created using 3-D computer animation, stereoscopic display, and highly complex data banks. Virtual reality can offer a sophisticated form of adaptive measurement. Emerging applications of virtual reality and simulation software include treatments for anxiety, pain management, claustrophobia, body image problems, phobias, and rehabilitation in brain injuries.

One research project has structured an interactive computer program to assess driving skills (Trilling, 2001). The ability to drive to work, visit a friend, or go to the grocery store signifies a high step toward independence for survivors of traumatic brain injury. It is also a public safety issue. The computer-based driving program allows the user to interact with a computer-generated environment that simulates a real-life environment including visual, auditory, and tactile cues. For clinicians, the virtual reality approach offers an objective measure of driving performance with an environment that can be modified into increasingly difficult scenarios such as navigating complicated detours or controlling the car in snow or rain.

Another rehabilitation project brings the benefits of virtual reality directly into the patient's home. It provides Internet-based modules addressing common daily activities so that individuals can practice using a microwave or an ATM banking machine. Working at home, the patient can practice as much as needed, building confidence and skill in a private setting (Lemaire & Green, 2002).

The development of a simulation tool is under way to allow for 3-D animation of joint motions (Van Sint Jan, 2000). The patient wears a "knee sock" with sensors that translate knee movements to a computer. This results in an image of a knee joint, complete with bone and ligaments, superimposed onto a 3-D image. The tool can translate MRI information into the image for clinicians evaluating patients who are undergoing physical rehabilitation.

Research is ongoing that pushes traditional treadmill therapy to another dimension by allowing stroke patients to take their first trip to the mall without ever leaving the security of a virtual reality world (Campbell, 2003). Patients wear a safety harness and train on motorized treadmills and virtual reality systems housed in rehabilitation centers. The treadmills are mounted on platforms that rock and tilt in synchrony with changing environmental animations displayed on large screens before the patient's eyes. This brings in the cognitive process and gives both the mind and the body a workout. Patients gain confidence, recover faster, and have fewer debilitating falls. Additionally, evaluation information is continuous and automatically recorded. The virtual reality therapy begins with simple movements for poststroke patients and gradually becomes more difficult over time. For example, an early scenario requires a patient to navigate an empty corridor. Eventually, in advanced scenarios, the patient may walk through crowded shopping malls or outside in rain or snow.

The field of bioengineering has been growing rapidly in the past few years (Sullivan, 2000). This expansion is due to many factors, including scientific and technological advancements, the increasing recognition of the role of interdisciplinary needs to solve complex biomedical problems, and the aging of the population with associated demands and costs. Bioengineers work on a broad range of things from molecules and genes to organs and whole-body systems, including drug delivery systems, biomechanics, gene systems and their organization, tissue engineering, organ replacement systems, artificial blood, and bioinstrumentation and physiologic monitoring. For example, every year more than 700,000 patients in the United States undergo highly invasive joint replacement surgery (Goho, 2003). Injectable tissue engineering is being researched. Already, researchers have developed a way to inject joints with specially designed mixtures of polymers, cells, and growth stimulators that solidify and form healthy tissue (Schultz et al., 2000). The new frontier lies in the use of stem cells.

Bioengineering is now in the early stages of a transforming technology, based on the intersection of biology and information, to understand the course and methods associated with health and disease processes. The hope is that by reprogramming the information processes that lead to and encourage disease and aging, we will have the ability to overcome these processes (Hausdorff et al., 1997).

New frontiers in distance education are developing rapidly, and convergences are leveraging these developments exponentially. Technology utilization will become a more natural and less onerous event. New Internet applications are developing that will enhance the ease with which information can be organized and will allow greater compatibility between technology and the person, such as with voice recognition software. Distance learning will become commonplace, and meetings of all kinds will routinely take place among geographically distant participants. The Internet will allow global access to assessment services and will provide an opportunity for resources to be leveraged internationally. Nano-biotechnology will bring tremendous advances in early detection of diseases and their treatment. We will see major advances for treating the loss or partial loss of auditory, visual, and sensory functions through the introduction of novel micro- and nano-engineered electronic devices (Freitas, 2002).

Nike has initiated the Oregon Project to deconstruct the elements of championship runners through the application of science, technology, and quantification (Tilin, 2002). We will begin to see patients who desire not only health and happiness but enhancements.

Smart technologies with wireless, wearable motherboards will facilitate applications ranging from infant monitoring (Shin et al., 2003) to sport feedback to geriatric care (Sands, Phinney, & Katz, 2000). Developments in neurosurgical and electrophysiological procedures are making feasible implants that hold promise of a more normal life for individuals with handicaps. Researchers at Emory University implanted a telecommunications device into a 53-year-old poststroke patient's brain that allowed the paralyzed man to move the cursor across the computer screen with his brain (Kennedy & Bakay, 1998).

We cannot predict *exactly* the future of science or technological innovation. We simply can't account for the multiplicity and interactivity of all the possible variables. Preparation for the future, however, does not require accurate prediction. Rather, it requires a foundation of knowledge upon which to base action, a capacity to learn from experience, close attention to what is going on in the present, and healthy and resilient institutions that can effectively promote healthy adaptation and respond to changes in a flexible and timely manner.

Part IV

Measurement in Kinesiology

Past, Present, and Future

Chapter 13

Measurement and Evaluation Council

Past, Present, and Future

Ted A. Baumgartner

It is an honor and a pleasure to talk about the Measurement and Evaluation Council. Just in case some of you don't know how the Measurement and Evaluation Council fits into an organizational structure, I am going to briefly address structure. Then I am going to address the Measurement and Evaluation Council in the past, present, and future. There are some things that are important to address, but they cut across several time periods.

Structure

The American Alliance for Health, Physical Education, Recreation and Dance (AAHPERD) is the professional association for many of us. One of the six associations of AAHPERD is the American Association for Active Lifestyles and Fitness (AAALF). The Measurement and Evaluation Council is one of the 12 councils of AAALF. Names of structures have changed over time, but I will use present-day names whenever possible.

Past

Measurement and Evaluation was a section within the AAHPERD structure from 1949 through 1972. The Measurement and Evaluation Section could have one meeting at the national convention. In 1973, Measurement and Evaluation became a council so it could have multiple meetings at conventions, hold meetings during the year, and sponsor projects. So, projects like the 1958 AAHPER Youth Fitness Test and the 1960s AAHPER Sports Skills Tests series were developed with little involvement of the Measurement and Evaluation Section. The development of the 1980 AAHPERD Health-Related Physical Fitness Test and the revision and expansion of the AAHPERD Sports Skills Tests series from 1984 to 1991 were conducted with considerable involvement of the Measurement and Evaluation Council.

There has been a series of Measurement and Evaluation Symposiums. The Measurement and Evaluation Council has been involved in all of these symposiums. This symposium we are attending is the 10th Measurement and Evaluation Symposium. Many of you may not be aware of the history of the Measurement and Evaluation Symposium series. Proceedings of most Measurement and Evaluation Symposiums were published. These Proceedings were very helpful as I prepared this presentation. I organized the first Measurement and Evaluation Symposium at Indiana University in 1975. Dale Mood organized the second symposium at the University of Colorado in 1977. Jim Morrow at the University of Houston and Jim Disch at Rice University organized the third symposium in 1980. Since then, a Measurement and Evaluation Symposium has been conducted every three or four years at the University of Northern Iowa, Louisiana State University, University of Wisconsin, University of Georgia, Oregon State University, and Cooper Institute in Dallas. I must tell you that organizing a Measurement and Evaluation Symposium is a lot of work, and both Jim Morrow and I have been involved in organizing two different symposiums. We should thank Weimo Zhu for organizing this symposium.

In the 1970s, many authors of measurement and evaluation books in physical education and exercise science had not concentrated in the measurement and evaluation area for their doctoral training. I don't think we can blame this on the Measurement and Evaluation Council. Later in my presentation I will mention measurement and evaluation books again.

Present

In talking about the present I am going to say "recently" without defining "recently." Maybe recently is from 1973, when Measurement and Evaluation became a council, to the present. The Measurement and Evaluation Council operating guidelines (1999) are quite informative as to the purpose of the council. "The purpose of this Council shall be to promote both practical and theoretical endeavors in the area of measurement and evaluation as it is applied in health, physical education, recreation, and dance." Some of the specific purposes stated in the guidelines are (a) to encourage and stimulate the use of new and accepted procedures related to the measurement of performance; (b) to stimulate the development of new techniques of evaluation of performance;

(c) to promote understanding of the use of evaluation procedures and tests; (d) to prepare and disseminate materials that will aid members; (e) to undertake or consult on the development, preparation, and revision of national measurement and evaluation projects of AAHPERD; (f) to maintain liaison with other structures of AAHPERD to provide advice and cooperation relative to any measurement and evaluation problems of these groups; (g) to promote the use of valid tests in health, physical education, recreation, and dance (HPERD); and (h) to encourage and support training and specialization in measurement and evaluation at the masters and doctoral level. I didn't realize that the Measurement and Evaluation Council is trying to do all of these things. Maybe measurement and evaluation specialists, as well as other people, need to be reminded of what the Measurement and Evaluation Council is supposed to do.

Recently measurement and evaluation specialists have authored most measurement and evaluation books in physical education and exercise science. I think this is due to more interest in measurement and evaluation and better-informed teachers of measurement and evaluation courses. Some of this must be credited to the Measurement and Evaluation Council for making measurement and evaluation more visible. The first editions of some of the leading measurement and evaluation books were published in the 1970s. If this can be attributed to the influence of the Measurement and Evaluation Council, should I discuss it in connection with the past or the present?

Measurement and Evaluation Symposiums have already been discussed. Presently they are still being conducted, but they began in the past and will probably continue in the future.

Individuals doing measurement research in physical education and exercise science develop new physical performance tests, measurement instruments, and measurement procedures. For simplicity of presentation, I am going to call these three things measurement techniques. Measurement researchers estimate objectivity, reliability, and validity for new measurement techniques. These estimates are good only for the population or populations used to obtain the estimates. Measurement researchers also estimate objectivity, reliability, and validity for old measurement techniques to be used with new populations. Measurement researchers may develop new uses of statistical techniques or new statistical techniques for use with physical performance data.

Measurement researchers need journals where they can publish their research. There are some journals that publish some measurement research dealing with physical performance, but these journals do not publish enough measurement research to accommodate all of the good measurement research dealing with physical performance. Educational measurement journals usually don't publish measurement research dealing with physical performance. The time came when a journal totally devoted to publishing measurement research dealing with physical performance was needed. Through the efforts of a number of measurement specialists in the Measurement and Evaluation Council and the efforts of AAALF, the journal *Measurement in Physical Education and Exercise Science* was started in 1997. This journal is a publication of the Measurement and Evaluation Council and AAALF. Any measurement research in physical education, exercise science, and related areas dealing with physical performance is appropriate for submission to *Measurement in Physical Education and Exercise Science*. In addition, book reviews, tutorials, teaching hints, and so on are published in *Measurement in Physical Education and Exercise Science* for teachers and practitioners.

Future

It takes many years for a journal to become established. Through the efforts of the Measurement and Evaluation Council, *Measurement in Physical Education and Exercise Science* can become a larger and more influential journal. However, for the next three to five years, measurement specialists must conduct more measurement research and publish it in *Measurement in Physical Education and Exercise Science*. In a relatively new journal, accepted manuscripts are published promptly because there is little backlog of accepted manuscripts. As a result of *Measurement in Physical Education and Exercise Science,* the quality and quantity of measurement research will increase. Also, measurement and evaluation practices will improve as a result of *Measurement in Physical Education and Exercise Science.*

There is a great need for measurement research devoted to the refinement of present-day measurement techniques, as well as for the development of new measurement techniques. In both cases, the purpose of the research must be to refine or develop the measurement techniques or to do both. There are too many measurement techniques in use today that were quickly and poorly developed for data collection in an experimental research study. There is a great need for research dealing with the measurement problems of the practitioner, as well as with highly sophisticated topics and techniques. All of the measurement research can't be high level and cutting edge. High-level research seldom addresses the problems of the practitioners and is seldom understood and appreciated by the practitioners. The individuals who are doing high-level and cutting-edge research should continue to conduct this research. However, all of us can do research to find answers to measurement problems that exist in everyday life.

There are many small, everyday measurement questions and problems that need to be researched. The push-up is commonly used today as a field-based test of arm and shoulder girdle strength and endurance. For the last few years I have conducted research dealing with the push-up as a test. Is a push-up test protocol that is good for college students good for elementary school children? Is a push-up test protocol that is good for male college students good for female college students and vice versa? If the down position for a push-up test is changed from the chest touching the floor to a 90° angle at the elbows, is the resulting push-up test a new test? If the change in the down position for the push-up test is a major change, then the push-up test is a new test. Validity and reliability evidence must be obtained for any new test before the test is used. I have seen the down position for the push-up test changed several times over the years, with seemingly no thought about the effects of the changes on reliability and validity for the test. Young children, females, and low-strength individuals used to do the bent-knee push-up test that is executed on the hands and knees rather than on the hands and toes as in the traditional push-up test. Today everyone does the push-up test on the hands and toes. Was there any evidence when the change was made, and is there any evidence now, that the push-up test on the hands and toes is valid for individuals who used to do the bent-knee push-up test?

This is just one example of the many measurement problems that measurement specialists and the Measurement and Evaluation Council could and should be investigating. The Measurement and Evaluation Council must take the lead in the refinement and development of measurement techniques. In terms of knowledge tests and other paper-and-pencil instruments, there are big test services like American College Testing and Educational Testing Service that develop and distribute tests and other measurement instruments. The Measurement and Evaluation Council could and maybe should become the equivalent of these companies for physical performance measurement and evaluation. The Measurement and Evaluation Council should sponsor some big measurement projects like developing and norming a nationally distributed physical fitness test or developing an authentic assessment program. Jo Safrit wrote in the February 1990 issue of *Pediatric Exercise and Sport* that probably no reputable test service in the country would publish any one of the existing fitness tests without additional evidence of the validity and reliability of the test battery, the validity and reliability of the test items, and so on. That is rather damning. Can you imagine a version of the SAT or ACT or GRE being quickly thrown together by a test service and distributed with hardly any validity, reliability, and objectivity evidence? Why do we do this with physical performance tests? Are things any better today than they were in 1990?

The Measurement and Evaluation Council needs to promote measurement more and take responsibility for what is happening in the measurement area. It needs to get more people interested and involved in measurement and active in the Measurement and Evaluation Council. Too few people are active in the Measurement and Evaluation Council. A few young measurement specialists seem to be inheriting all of the leadership responsibilities in the council. There seems to be a shortage of measurement people to review measurement research manuscripts submitted to journals. Two to four people are training measurement specialists in physical education and exercise science. With the exception of one person, the rest of the people training measurement specialists are close to retirement. To my knowledge, the Measurement and Evaluation Council has never been concerned with the training of measurement specialists, although encouraging and supporting training of measurement specialists is one of the purposes of the Measurement and Evaluation Council. As people training measurement specialists retire, the Measurement and Evaluation Council needs to encourage other people to start training measurement specialists. It does not seem to me that the Measurement and Evaluation Council has been concerned with who will be the next editor of *Measurement in Physical Education and Exercise Science* and the identification of possible applicants. It would be embarrassing to have only one person training measurement specialists in physical education and exercise science or no new editor of *Measurement in Physical Education and Exercise Science*. The Measurement and Evaluation Council needs to take some responsibility and become proactive in regard to these issues.

Probably the Measurement and Evaluation Council is going to have to expand its view as to areas in which measurement and evaluation occurs that are of interest to or are the responsibility of the Measurement and Evaluation Council. Notice that in the operating guidelines for the Measurement and Evaluation Council the areas of responsibility are health, physical education, recreation, and dance. We can't continue to limit ourselves to physical education and exercise science or HPERD. We can no longer just be interested in measurement of physical education skills and fitness.

There are athletic trainers with unique measurement situations and problems. Presently at the University of Georgia, the athletic training faculty and graduate students in the Department of Exercise Science are researching the measurement of concussions. They need a measurement technique to determine if an athlete has a concussion and when an athlete can return to competition after a concussion. The adapted physical education people have some unique measurement problems. The Measurement and Evaluation Council worked with the adapted physical education people years ago and is starting to work with the Adapted Physical Activities Council in terms of measurement problems and measurement technique development. There are many problems in measuring and evaluating the physical performance of the elderly. Good laboratory- and field-based measurement techniques are needed. This is why we are having this symposium.

In many cases, to develop good measurement techniques requires a committee who are willing to devote a considerable amount of time to the project. Some members of the committee must be content specialists, and some other members of the committee must be measurement and evaluation specialists. The content specialists make sure the measurement and evaluation specialists don't make any huge mistakes in designing the measurement technique, and the measurement and evaluation specialists make sure the content specialists follow sound measurement and evaluation principles in developing the measurement technique. The Measurement and Evaluation Council members must take the initiative to start these projects but also be available if asked to join a project initiated by individuals in a content area.

Who are the members of the Measurement and Evaluation Council? Are all of the members measurement and evaluation specialists or teachers of measurement and evaluation courses at the college/university level? Where do people with an interest in measurement and evaluation from a researcher or practitioner standpoint fit into the Measurement and Evaluation Council? Fourteen hundred seventy-nine AAHPERD members checked Measurement and Evaluation Council on their membership application as one of their areas of interest. I can't believe that all of these 1,479 people have more interest and ability in the measurement and evaluation area than the rest of the AAHPERD membership. In the future, Measurement and Evaluation Council officers must take the list of individuals who checked Measurement and Evaluation Council on their membership application and determine what these individuals want from the Measurement and Evaluation Council and what these individuals want to contribute to the Measurement and Evaluation Council. There are many people on the list who are retired or eminent in their area and active in their council. I doubt that these people will be active in Measurement and Evaluation Council programs.

How do we develop measurement techniques for populations other than college students or people on campus? How do we develop norms for these measurement techniques? Either we send people out to test these populations or we convince the people working with these populations to do the testing. In the past, neither approach has been very effective. We must use the individuals on the Measurement and Evaluation Council list who want to contribute to Measurement and Evaluation Council projects to do some of the testing. I suspect there are many individuals at small colleges and public schools who are not measurement specialists but have an interest in measurement and would be glad to help with Measurement and Evaluation Council projects that are practical and meaningful to them. In terms of dramatically improving measurement and evaluation understanding and practices, and substantially increasing the number of good measurement techniques for a variety of populations and both genders, the sky is the limit if we will utilize many of the 1,479 individuals who checked Measurement and Evaluation Council on their membership application.

Chapter 14

The Changing Face of the Measurement Specialist in Kinesiology

Stephen Silverman, EdD

At the 10th Measurement and Evaluation Symposium I delivered the banquet speech, and this paper is a summary of that talk. Over the years the person doing the banquet speech has taken different approaches in completing the task. Some have delivered an entertaining talk that poked fun at the content of the symposium and provided a light-hearted ending to the gathering. Other speakers have discussed measurement-related issues to stimulate discussion or to drive home a point to those attending the symposium. Still others have combined the two approaches. In my experience each talk has added to the meeting and has ended it on a positive note.

When I was asked to speak at the banquet, I thought it would be hard for me to be purely entertaining and comment on what had occurred at the symposium. As I considered a title, I had been thinking a great deal of how the symposiums had changed in the 20 years I had been attending them. I had been thinking about who had attended and where the new measurement specialists were being trained and areas in which they were doing scholarship. Although I wanted the talk to start out light-hearted, I thought it would be good to think further about these topics and organize my talk around them. As with the talk, the purpose of this paper is to discuss changes in the kinesiology measurement and evaluation (M&E) field and look toward where we are going in the future. I will discuss how the field has changed over the past 30 years, then present issues I believe are important for us to consider, and end with a conclusion.

Coming Clean

Those of us who read qualitative research articles have seen a very positive trend in which the author of the article presents her or his biography to help readers place the results and conclusions in context. I believe this is important and along with my coauthors (Locke, Silverman, & Spirduso, 2004) have suggested that we should expect this when reading articles about studies in which the individual is the source of data collection and analysis. While I am not a qualitative researcher and this is not a research paper, it is appropriate for me to provide a little background to help readers understand my perspective.

In reality I am not a measurement scholar. My primary area is pedagogy and, for a number of years, I have conducted research on teaching in physical education. I also have had strong interests and experiences in areas that are often considered the purview of kinesiology M&E specialists. Some of my research has focused on how to collect data in field settings—mostly reliability and validity papers so my collaborators and I can collect data in physical education classes. This research has been published in kinesiology and educational measurement journals. I have coauthored two research methods texts that have been published in multiple editions. I also have taught research methods and measurement classes every year since I have had a professorial appointment. I like to think of myself as a pedagogy scholar who has strong measurement interests. You, the reader, however, must determine whether I am an insider commenting on these issues or someone who is outside his area of expertise.

Changes in the Measurement and Evaluation Field

The field of M&E has seen many changes over the past 30 years. These changes are a reflection of student interests, departmental needs, the increased focus on scholarship in our field, and changes that have occurred in other social science areas and in the field of kinesiology. In this section, I will discuss these changes from the 1970s until the present.

The Young Face of the Measurement Specialist (Circa 1970s)

As Ted Baumgartner (2003) noted on the opening evening of the symposium, measurement specialists in the 1970s looked different than they do now. I was an undergraduate physical education major in the early to mid-1970s and took my first M&E class then. Those who were teaching M&E may or may not have been trained as measurement specialists. Typically, they taught undergraduate classes in M&E for students who had the goal of becoming physical education teachers. In many cases, they also taught graduate classes in measurement, research methods, and statistics. Some taught one or more classes in another area within the field (e.g., exercise physiology or motor behavior). Few of those whom we would consider measurement specialists during that time had a research program that focused on or had a strong component related to measurement.

The Middle-Aged Face of the Measurement Specialist (Circa 1980-1990s)

The period of the 1980s and 1990s provided the field with many more specialists in M&E. This mirrored the increase in specialists in other areas within kinesiology (e.g., pedagogy, sport sociology, sport and exercise psychology), and a number of those in the other areas had strong support in M&E. Measurement specialists who took positions during this time taught undergraduate classes in M&E—for students who wanted to be teachers and for those who aspired to other careers within kinesiology. In some institutions, an M&E class was no longer a requirement for graduation, and classes may have been segregated by the career choices students were making. Measurement and evaluation specialists often taught graduate classes in measurement and in research methods and sometimes in statistics. In addition, a number of M&E textbooks appeared during this time period.

Along with teaching responsibilities, measurement specialists were conducting more focused and productive research programs. This research was sometimes applied, focusing on the application of measurement or the ways in which measurement is used in professional settings (e.g., Hensley et al., 1989; Safrit & Wood, 1986), and sometimes theoretical (e.g., Godbout & Schutz, 1983; Looney & Spray, 1992; Safrit et al., 1992), often focusing on the reliability and validity of scores from different tests and assessments (e.g., Jackson et al., 1996; Smith et al., 1995). The great increase in scholarship helped advance M&E and the other areas within kinesiology, since once measurement specialists published papers the techniques and knowledge often were used by those in other areas.

The focus on research and scholarship resulted in many more outlets for M&E research. A number of content-specific journals (e.g., *Journal of Exercise and Sport Psychology* and *Journal of Teaching in Physical Education*) appeared that welcomed measurement-related papers. *Measurement in Physical Education and Exercise Science* was founded in 1997 and quickly became an outlet for measurement specialists. Broader kinesiology journals (e.g., *Research Quarterly for Exercise and Sport* and *Journal of Human Movement Science*) saw a greater frequency of measurement papers. As a sign that kinesiology measurement specialists also were influencing the broader area of measurement, papers were published in social science measurement journals (e.g., *Journal of Applied Measurement* and *Educational and Psychological Measurement*). The amount of research being published was impressive and contributed greatly to research advances in kinesiology.

The Current Face of the Measurement Specialist (Circa Now)

While the field has moved forward, middle age and more advanced age have brought many changes. There still are many specialists who were trained in

M&E and others who have a strong background in M&E. The participants at the symposium represented both M&E specialists and those of us with specialties in other areas who devote some of our scholarly interest to measurement issues. It must be noted, however, that both the number and the proportion of those who were trained as measurement specialists are getting smaller. Many of the distinguished senior measurement scholars have retired or will retire soon.

Faculty members with a measurement interest do what they were doing in the 1980s and 1990s—teaching, research, and publishing measurement-related papers. Many of our colleagues are working with others and bring the technical expertise of the measurement specialist to research in other areas. These articles are being published in many journals. Although still in many ways thriving, the measurement specialist is getting older.

Issues and Questions for the Future

Having presented my perceptions of the M&E field for the past 30 or so years, I will focus on issues and questions that I believe will influence the field over the coming years. I have selected these issues based on observations from colleagues and my own work in the field. The questions we should be thinking about are the following: (a) Who teaches M&E classes, and how are they taught? (b) How do we help students learn about research methods? (c) Where will measurement specialists be employed? (d) How will kinesiology faculty be trained in M&E? and (e) Who will publish measurement papers?

Who Teaches, and How Are Measurement and Evaluation Classes Taught?

This question is important because undergraduate and graduate students at many colleges and universities take measurement classes. I would venture that attendees of the symposium and anyone who chooses to read the Proceedings believes that a background in measurement enhances the knowledge and subsequent professional practice of those entering the field. From knowledge about evaluating the most appropriate assessment for measuring range of movement in an athletic training setting, or about assessing student progress in schools, to an understanding of issues related to measurement of fitness and physical activity in children, we would hope persons with degrees in kinesiology would understand these issues better than others outside kinesiology. Measurement and evaluation can enhance what our graduates do—and help provide a basis for thoughtful professional practice.

It is hard to determine if students are receiving a good grounding in the practical or theoretical aspects of measurement. I spent some time on the Internet looking at those who teach measurement-related classes at colleges and universities and the classes that are offered in the area in which I live (the tristate area around New York City). While this area is different from others in many ways, I think those who are teaching measurement classes and the focus of those classes are similar to those in much of the rest of the country. At many institutions, the person teaching measurement classes is not trained in measurement (see Baumgartner & Safrit, 2003, for a genealogy of measurement specialists) nor, in my estimation, does the person have a strong measurement background. The classes that are being taught appear to be increasingly segregated by the concentration the student is pursuing (i.e., assessment courses for teacher education students and other courses for exercise science students), and online syllabi did not reflect much of an emphasis on measurement theory. Obviously, who and what is taught are intertwined. The question remains, however, whether undergraduates will get a broad background in measurement and whether they will get any theory if the faculty members teaching the courses do not have a strong measurement background.

How Do We Help Students Learn About Research Methods?

When students take research methods classes we would like them to understand research, stay current with the research and use it in their professional practice, and, in some cases, be able to design and conduct research. As I have noted before (Silverman, 2002), it is very difficult to do this in one graduate class, and many graduate research methods classes have so many goals (Silverman & Keating, 2002) that it becomes difficult for students to learn or, more particularly, to develop good attitudes and an appreciation for research.

Who teaches research methods probably will determine how they are taught and the focus of the class. There are a number of interrelated questions that we should consider when we think about research methods classes. Without additional work or training, can researchers who are skilled in methods in their area teach a class that has students from many areas within kinesiology? How will instructors integrate new research methods (e.g., methods for determining reliability and validity of scores or qualitative methods) into their classes? Will instructors have the knowledge and the pedagogical content knowledge to teach students who will be users of research or who hope to go on to careers in which they facilitate research use by others (e.g., instructional supervisors in schools or managers of exercise facilities)?

The answers to these questions will help determine the research competence of the next generation of kinesiology graduates. If these graduates take research courses that do not change with the times or do not help them reach their professional goals, research literacy will decrease. In my experience and from an empirical analysis (Silverman & Keating, 2002), many research methods classes reflect the research area of the instructor and have a heavy emphasis on statistics. With apologies to Jo Safrit, who once did something similar in a presentation at an AAHPERD meeting and for which I cannot find the citation, we should remember that *RM ≠ M&E ≠ statistics*.

Classes in research methods require specialists who tailor them to the students in the class. If we want our students to be able to read and understand research, we cannot assume that all students will do research and that by learning statistics they will be able to understand and critique research methods used in papers in their interest area.

Where Will Measurement Specialists Be Employed?

If we want faculty members with strong backgrounds in M&E to teach measurement and research methods courses, departments need to hire those individuals. In addition, students need to be graduating in the area who are interested in assuming these positions. In both these cases, things do not seem promising. Woods, Goc Karp, and Feltz (2003) present an analysis of the college and university job market in departments of kinesiology and physical education. Their data suggest that there are very few M&E positions in higher education. In fact, less than 10 M&E positions were advertised over a four-year period. Only one of these positions was at a research-intensive or a research-extensive university where the new faculty member may have an opportunity to work with doctoral students wishing to study M&E and take future positions. It is hard to tell whether the job market influences who studies M&E or whether the lack of candidates for positions influences how departments structure positions. In any case, those teaching measurement in the future are likely to be trained in another area. This, of course, can be positive or negative—depending on the background of the person teaching these classes and his or her attempts to develop classes that meet the needs of the given students and departments.

How Will Kinesiology Faculty Be Trained in Measurement and Evaluation?

An issue related to employment and job prospects is how both measurement specialists and other faculty in kinesiology will be trained in measurement and related fields. In many areas within kinesiology, greater sophistication is now required to conduct research than in the past. For example, in my area of pedagogy, researchers need to provide evidence that the scores from measurements are reliable and valid. The methods used to provide evidence today are far different from those when many of us were in graduate school. If we are to have good researchers, graduate students must get a good background and have the skills, knowledge, dispositions, and resources to keep up with advances in the field.

The obvious way this might occur is if measurement specialists in kinesiology departments provided advanced training to graduate students and to faculty. In order for this to happen there will need to be measurement specialists in kinesiology departments who are specifically trained in these areas. I looked at the Web pages of 59 institutions offering doctoral degrees in kinesiology or a related area. These, as best as I can tell from a list assembled by the American Academy of Kinesiology and Physical Education and from a Google search, represent all departments that offer the doctorate in the United States. Only two of these departments (University of Georgia and University of Illinois at

Urbana-Champaign) currently have a program in M&E that leads to a terminal degree. Many of the universities from which current measurement specialists graduated no longer offer a specialization. As M&E faculty have retired, so have departments retired M&E specializations. Given the job market and the two PhD programs, it is unlikely that most departments will have a measurement specialist who is trained in that area.

One way in which many departments are dealing with the need for measurement expertise is through kinesiology faculty members in other areas who have a background in measurement-related content. At this symposium, many of the speakers and participants do not have a doctoral specialization in measurement but have strong supporting work and have worked throughout their careers to keep up-to-date on measurement issues and techniques that are important to their research and scholarship. Many have also conducted measurement research in their specialty areas. These people are excellent resources for our graduate students and for other faculty and students in their departments. The fact that more faculty with measurement expertise do not look like those who were measurement specialists in the past suggests that the face of the measurement specialist, in many kinesiology departments, has greatly changed.

Since at many universities there is not someone designated as a measurement specialist who formally works to teach classes and help educate colleagues and students about advances in measurement, most departments utilize other resources on their campuses for courses and consulting. For example, when I was a faculty member at Illinois, my PhD students took classes in test theory from Rod McDonald, a world-renowned psychometrician. He also served as a member of a number of my students' dissertation committees. As many of the symposium participants learned from the preconference workshop that Rod presented, he is an excellent teacher, and most departments would be hard pressed to find anyone better to teach that class and to provide students with measurement expertise. Many kinesiology departments have shifted the theoretical and technical classes in measurement and in statistics to their colleagues in educational psychology or in psychology. This is a valuable and efficient way to develop expertise in the next generation of kinesiology scholars. Of course, as in everything graduate advisors do, we should make certain that the classes meet the needs of our students and that the faculty members teaching the kinesiology students are receptive to questions and to the problems they are likely to investigate.

Who Will Publish Measurement Papers?

This final question will determine whether or not measurement will be a visible part of our scholarly fabric in the future. There is ample evidence to suggest that specialists in other areas of kinesiology will continue to publish papers about M&E (e.g., Keating, 2003) and reliability and validity (e.g., Kulinna, Cothran, & Regualos, 2003). These papers are important to our professional dialogue and to conducting good research. As was noted earlier, kinesiology faculty and graduate students are developing measurement skills that permit and encourage this type of scholarship. The question, however, is whether we will see papers published that advance measurement theory in kinesiology. It is unlikely that those scholars in departments of educational psychology or psychology will have an interest in the field of kinesmetrics, to use the term Weimo Zhu has coined. If M&E is to continue to advance, there must be a sufficient core of scholars doing research and publishing in the area. A few isolated scholars will not advance our theoretical understanding as will a group of scholars who share information and create a scholarly community. More expertise is good. We should look for it in the places we have in the past—and in places where it is now located.

Conclusion

As Yogi Berra once said, "It's tough to make predictions, especially about the future." I believe, however, that we are seeing the face of the kinesiology measurement specialist mature and change. The change is in a specific direction. As Yogi also said, "It's like déjà vu all over again." Thirty years ago most of the people teaching M&E were not measurement specialists, and that is true again today. There are only a few positions advertised each year and few places where one could get an M&E PhD. On the other hand, the mature face of our field reflects a strong but smaller group of measurement specialists and a growing group of kinesiology faculty members who have a strong interest in measurement.

The face of the measurement specialist has changed—and perhaps changed back. As a group we should continue to integrate good measurement into our research and work to ensure that the next generation of scholars has a good measurement background. This will help us grow and keep a smile on our collective face.

Appendix

*The 10th Measurement and
Evaluation Symposium Program*

10TH MEASUREMENT AND EVALUATION SYMPOSIUM

AAALF, ACSM, Human Kinetics Publishers and
The University of Illinois at Urbana-Champaign Present:

The 10th Measurement and Evaluation Symposium:

Measurement Issues and Challenges in Aging Research

October 16-18, 2003 • University of Illinois
Urbana-Champaign, Illinois, USA

Advance Program

Dear Colleagues,

It is my great pleasure to invite you to participate in the 10th Measurement and Evaluation Symposium: Measurement Issues and Challenges in Aging Research.

Established in 1949, the Measurement and Evaluation (M&E) Council is one of 12 councils/societies in the American Association for Active Lifestyles and Fitness (AAALF), which is one of six associations in AAHPERD. The M&E Council holds its symposium every three or four years, with a focus on measurement and evaluation, statistical, and research design issues in the field of Kinesiology.

In the past, only measurement specialists, or researchers who are interested in measurement and statistics, came to the symposium. When Professor Terry Wood at the Oregon State University organized the 8th M&E Symposium in 1996, he invited leading researchers in six sub disciplines to address measurement issues/challenges in these disciplines. Measurement specialists were also invited to react to these issues and challenges. This "expert vs. measurement specialist" interaction format created a unique forum for outlining and debating methodological and philosophical challenges when quantifying human behavior. Integrating with the Cooper Institute Conference series and organized by Drs. James Morrow, Jr. and Steven Blair, the 9th M&E Symposium was another great success. Focusing on the theme "Measurement of Physical Activity," many leading researchers in the field were invited to address the key issues and challenges in measuring physical activity. The symposium drew more than 200 researchers from all over the world, and the research papers presented were published in a special supplement of the Research Quarterly for Exercise and Sport.

When the M&E Council was planning its 10th symposium, it contacted me for the possibility to hold the symposium at the University of Illinois at Urbana-Champaign (UIUC). Considering the fast-growing interest in aging research in the field and with the full support of Professor Wojtek Chodzko-Zajko, the head of the Department of Kinesiology at UIUC, we have selected the "Measurement Issues and Challenges in Aging Research" as the theme for the symposium. AAHPERD, ACSM, and Human Kinetics (HK) have agreed to sponsor the conference.

I am extremely delighted that so many world leaders in aging and measurement have agreed to speak at the symposium. We will keep the "expert vs. measurement specialist" format with a scientific presentation on a particular issue in aging first, followed by methodological considerations. Other sessions will focus on new, unexplored research areas and issues in aging research. In addition, we have, for the first time, included two pre conference workshops to provide an update on two major measurement techniques: Structural equation modeling and longitudinal data analysis. Participations are also encouraged to submit the results of their own research studies for the poster sessions. The final session of the symposium will summarize and coalesce the issues and interests raised in the symposium and bring closure and recommendations. The papers presented will be published in a proceeding book and every conference participant will receive a free copy of the book. The abstracts presented will be published in the journal of the Measurement in Physical Education and Exercise Science.

The 10th M&E Symposium will be an exciting and intellectually stimulating gathering for both aging and measurement researchers, and I am looking forward to seeing you in October at UIUC campus!

Weimo Zhu, Ph.D.
Chair, Organization and Scientific Program Committees

Sponsors

Faculty

Ted A. Baumgartner, Ph.D., University of Georgia

Mr. Ping Zhang Chen, Shanghai Qi-Gong Research Institute, China

Uriel Cohen, D. Arch., University of Wisconsin, Milwaukee

Wojtek Chodzko-Zajko, Ph.D.,
University of Illinois at Urbana-Champaign

Leigh Jerome, Pacific Telhealth and Technology Hui

Marilyn Looney, P.E.D., Department of Kinesiology,
Northern Illinois University

Roderick P. McDonald, Ph.D., M.Sc., D.Sc., University of
Illinois at Urbana-Champaign

Edward McAuley, Ph.D., University of Illinois,
Urbana-Champaign

James R. Morrow, Jr., Ph.D., University of North Texas

Marcia Ory, Ph.D., Texas A & M University

Patricia Patterson, Ph.D., San Diego State University

Miriam E. Nelson, Ph.D., Tufts University

Tuomo Rankinen, Ph.D., Pennington Biomedical
Research Center

Roberta Rikli, Ph.D., California State University, Fullerton

James Rimmer , Ph.D., University of Illinois at Chicago

David Rowe, Ph.D., East Carolina University

Margaret Jo Safrit, Ph.D., American University

Robert Schutz, Ph.D., University of British Columbia

Roy J. Shephard, M.D., Ph.D., D.P.E., FACSM,
University of Toronto

Stephen Silverman, Columbia University

Waneen W. Spirduso, Ed.D., University of Texas

George Stelmach, Ph.D., Arizona State University

Jerry Thomas, Iowa State University

Renwei Wang, Shanghai Physical Culture Institute, China

John E. Ware, Jr., Ph.D., QualityMetric Incorporated

John B. Willett, Ph.D., Harvard Graduate School

Terry Wood, Ph.D., Oregon State University

Weimo Zhu, Ph.D., University of Illinois at
Urbana-Champaign

*Additional information about the faculty can be found on the ACSM Web Site, www.acsm.org.

156

Program

Thursday, October 16, 2003

8:30 a.m.-4 p.m.	Pre-Conference Workshop
	Morning Topic: SEM: Progress and Future Direction Roderick P. McDonald, University of Illinois at Urbana-Champaign
	Introduced by Weimo Zhu,
	University of Illinois at Urbana-Champaign
	Afternoon Topic: Improving the Analysis of Longitudinal Data
	John B. Willet,
	Harvard Graduate School
	Introduced by Hal Morris,
	Indiana University
5 p.m.-9:30 p.m.	Conference Begins
5-6:30 p.m.	Registration
6:30-6:40 p.m.	Opening/Acknowledgments
	Weimo Zhu
	University of Illinois at Urbana-Champaign
6:40-6:55 p.m.	Welcome
	Wojtek Chodzko-Zajko
	Head, Department of Kinesiology
	University of Illinois at Urbana-Champaign
	Tanya M. Gallagher
	Dean, The College of Applied Life Studies
	University of Illinois at Urbana-Champaign
	Chancellor/Provost
6:55-7:25 p.m.	Measurement Evaluation Council: Past, Present and Future
	Margaret Jo Safrit, American University
	Introduced by Ted A. Baumgartner,
	University of Georgia
7:25-8:10 p.m.	Measuring the Physical Domain of Older Adults: From Function to Consequences
	Waneen W. Spirduso, University of Texas
	Introduced by Wojtek Chodzko-Zajko at Urbana-Champaign
8:10-8:15 p.m.	Where do we go from here?
	Weimo Zhu, University of Illinois at
	University of Illinois at Urbana-Champaign
8:15-9:15 p.m.	Reception

Friday, October 17, 2003

8 a.m.-5 p.m.	Abstract Posters Displayed
8-11 a.m.	Section 1
	Presider: Jim Disch,
	Rice University
8-8:40 a.m.	Psychological Well-being and Quality of Life
	Edward McAuley, University of Illinois

8:40-9:10 a.m.	Measurement Specialist 1
	David Rowe, East Carolina University
9:10-9:20 a.m.	Break
9:20-10 a.m.	Environment and Physical Activity of Older Adults
	Uriel Cohen, University of Wisconsin-Milwaukee
10-10:30 a.m.	Measurement Specialist 2
	James R. Morrow, Jr., University of North Texas
10:30-11 a.m.	Discussion of Section 1
11 a.m.-1 p.m.	Poster Sessions and Lunch
12-1 p.m.	Authors will be present
	Presider: Jim Disch, Rice University
1-4 p.m.	Section 2
	Presider: Andrew S. Jackson,
	University of Houston
1-1:40 p.m.	Assessment of Physical Frailty and Disability
	Miriam E. Nelson, Tufts University
1:40-2:10 p.m.	Measurement Specialist 3
	Roberta Rikli,
	California State University-Fullerton
2:10-2:20 p.m.	Break
2:20-3 p.m.	Exercise Dose-Response Effects on Older Adults
	Roy J. Shephard, University of Toronto
3-3:40 p.m.	Human Gene Map, Physical Activity and Aging
	Tuomo Rankinen, Pennington Biomedical Research Center
3:40-4 p.m.	Discussions of Section 2
4-4:15 p.m.	Break
4:15-5:45 p.m.	Section 3
	Presider: Weimo Zhu, University of Illinois at Urbana-Champaign
4:15-5:30 p.m.	Measurement Challenges in Alternative Medicine: "Qi" and Aging
	Ping-Zhang Chen, Shanghai Qi-Gong Research Institute, China
	Renwei Wang, Shanghai Physical Culture Institute, China
	Shan S. Wong, National Center for Complementary and Alternative Medicine, NIH
	(an interactive session; translated by Weimo Zhu)
5:30-5:45 p.m.	Q/A and Discussion
	Dinner on your own

157

Saturday, October 18, 2003

8 a.m.-5 p.m.	Abstract Posters Displayed
8-11 a.m.	**Section 4** *Presider: Matthew T. Mahar, East Carolina University*
8-8:40 a.m.	Assessment Challenges in Persons with Disabilities *James Rimmer, University of Illinois at Chicago*
8:40-9:10 a.m.	Measurement Specialist 5 *Marilyn Looney, Northern Illinois University*
9:10-9:20 a.m.	Break
9:20-10 a.m.	Motor Control and Cognition Research in Elderly *George Stelmach, Arizona State University*
10-10:30 a.m.	Measurement Specialist 6 *Robert Schutz, University of British Columbia*
10:30-11 a.m.	Discussion of Section 4
11 a.m.-1 p.m.	Poster Sessions and Lunch
12-1 p.m.	Authors will be present *Presider: Yuanlong Liu, Western Michigan University*
1-4 p.m.	**Section 5** *Presider: Dale Mood, University of Colorado*
1-1:40 p.m.	Translating Research to Practice: Real World Issues in Moving from Efficacy to Effectiveness Research *Marcia Ory, Texas A&M University*
1:40-2:10 p.m.	Measurement Specialist 8 *Patricia Patterson, San Diego State University*
2:10-2:20 p.m.	Break
2:20-3 p.m.	Emergent Technologies and Remote Clinical Assessment *Leigh W. Jerome, Pacific Telhealth and Technology Hui*
3-3:40 p.m.	Online Computerized Adaptive Assessment of Disease-Specific Impact *John E. Ware, Jr., Quality Metric, Inc.*
3:40-4 p.m.	Discussion of Section 5
4-4:15 p.m.	Break
4:15-5:15 p.m.	**Section 6** *Presider: Sang-Jo Kang, Korean National University of Physical Education* Discussion of the Aging and Measurement Issues Raised at Symposium *Terry Wood, Oregon State University*
5:15-7:30 p.m.	Relaxation/Recreation Time
7:30-9:30 p.m.	Closing Banquet *Stephen Silverman, Columbia University* *Introduction, Jerry Thomas, Iowa State University*

Conference organizers retain the right to make programmatic adjustments, while maintaining the integrity of the educational offering

Sponsors

This symposium is sponsored and managed by the following: The American Association for Active Lifestyles (AAALF); the American College of Sports Medicine (ACSM); The University of Illinois, Urbana-Champaign; and Human Kinetics Publishers. AAALF and ACSM are national membership organizations. Contact Information: AAALF, American Alliance for Health, Physical Education, Recreation & Dance, 1900 Association Drive, Reston, VA, USA, Tel.: 1-800-213-7193, ext. 430, Web Site: www.aaphperd.org; ACSM, 401 W. Michigan St., Indianapolis, IN, 46202-3233, USA, Tel.: (317) 637-9200, www.acsm.org.

Target Audience

Aging researchers, measurement specialists, disability researchers, exercise specialists, fitness leaders, athletic trainers, personal trainers, alternative medicine practitioners, health care providers, physical therapists, physical educators, adapted PE specialists, and health/aging policy makers.

ACSM CEC's

Endorsement is currently being sought through the American College of Sports Medicine's Professional Education Committee.

Hotel Accommodations

Accommodations are available at discounted rates at two hotels. The Hampton Inn is located near campus, 1200 West University Avenue. Rooms are $60 U.S. dollars single or double, and can be reserved by calling Tel.: (217) 337-1100. Rooms are also available at the Holiday Inn Hotel and Conference Center, 1001 W. Killarney. Bus transportation will be available to take Holiday Inn guests to and from the Campus. Rooms are $72 U.S. dollars single and double and can be reserved by calling Tel.: (217) 328-7900. Reservations must be made by September 15, 2003 to access discounts, and are available on a first-come, first-served basis.

Traveling to Champaign-Urbana

Champaign-Urbana is located in the east-central part of Illinois at the intersection of Interstate Highways I-57, I-74 and I-72. Urbana-Champaign is accessible by air, rail, auto and bus. The Willard Airport in Champaign is approximately 10 minutes from the campus by taxi. Major carriers servicing the airport include American Airlines, American Connection and Northwest. Shuttle service is available from the airport to Campus from $10-$15.

Mid-October will yield beautiful, fall colors and temperatures will be generally mild with a range of 50°-70°. The University of Illinois campus will be bustling with fall activities including sporting events, special activities, and the typical goings-on of a Big Ten campus.

**AAALF, ACSM, Human Kinetics Publishers and
The University of Illinois at Urbana-Champaign Present:**
The 10th Measurement and Evaluation Symposium:
Measurement Issues and Challenges in Aging Research

10TH MEASUREMENT AND EVALUATION SYMPOSIUM **October 16-18, 2003 • University of Illinois, Urbana-Champaign, Illinois, USA**

CALL FOR ABSTRACTS

AN INVITATION
Abstracts for presentation via poster format are currently being accepted as part of the 10th Measurement and Evaluation Symposium: Measurement Issues and Challenges in Aging Research. You are invited to submit abstracts for publication and display. All abstract submitters are notified about acceptance, and should plan to register for the meeting. Do not submit an abstract if you do not plan to attend the meeting.

CONFERENCE DESCRIPTION AND INFORMATION
Established in 1949, the Measurement and Evaluation (M&E) Council is one of the established AAHPERD councils with a long and rich history. This 2003 co-sponsored event will focus on measurement issues and challenges in aging research, including a one-day pre-conference workshop on structural equation modeling and longitudinal study. The conference will take place on the University of Illinois Campus in Urbana-Champaign, Illinois. Nearby hotel accommodations have been secured. To register for the meeting, visit one of the following Web Sites: www.acsm.org; www.aahperd.org/aaalf.

RULES FOR SUBMISSION
Each person is permitted to submit and be first author on only one abstract, but may co-author any number of abstracts. The first named author must present the abstract. There is no abstract submission fee. **Submissions must be made no later than 5 p.m. CST on August 4, 2003.**

Conference organizers will be accepting abstracts electronically, via web site, exclusively. Electronic submission provides many enhancements including: the ability to view abstracts in their entirety online before the conference using an online itinerary program; the ability to search the abstract database by author, key-word, title, or institution. In addition, your notification regarding receipt of your material will be instantaneous. Additional instructions regarding the submission appear on the ACSM Web Site: www.acsm.org.

NOTIFICATION
You will be notified regarding acceptance or rejection of your abstract no later than August 25, 2003, but you may receive communication, specific to your submission, earlier. Submissions will be peer reviewed. An e-mail address is required in order to submit on-line.

PRESENTATION OF ABSTRACTS
Posters will be presented on Friday and Saturday, during the mid-day break (11 a.m.-1 p.m.) Information on the presentation schedule will be included with notification regarding your submission. Poster board dimensions are 8' (horizontal) x 4' (vertical). Time for placement is permitted prior to the actual sessions.

PREPARING THE ABSTRACT
Abstracts will be accepted in one of three categories: *measurement and aging*; *measurement*; and *aging*. Accepted abstracts will be published in the journal *Measurement in Physical Education and Exercise Science*. Abstracts are limited to 2,300 characters and must be no less than 1,300 characters. If including a table, chart, or graph, character reduction will be approximately 400 characters per exhibit. Be prepared to address any commercial disclosures. Grant funding information may be included.

Additional Instructions:
<u>Title</u>: The title should be brief (limited to 15 words).

<u>Authors</u>: Type the presenting author first. Include highest degree.

<u>Affiliations</u>: Type the affiliation where the research was done. Give only the name of the institution, city, state or province, country, and if you desire, the e-mail address of the first author.

<u>Text</u>: The abstract must be informative, including a statement of the study's specific purpose, methods, summary of results, and conclusion statement. It is unsatisfactory to state, "The results will be discussed." Abstracts must be written and submitted in English. Use standard abbreviations, symbols, and punctuation as typical of educational journals. When using abbreviations in the body of the abstract, spell out in full, the first mention, followed by further abbreviations in parentheses. Animal studies must comply with the National Institutes of Health (NIH) guidelines regarding use of animals. Once you have read all of the instructions, prepared the abstract text and charts, and are ready to submit, go to www.acsm.org. Instructions on the site will guide you through the electronic submission process.

WITHDRAWALS
Withdrawals must be made in writing. Send a copy of the abstract with a letter of withdrawal to:
Weimo Zhu, ACSM, 401 W. Michigan St., Indianapolis, IN 46202-3233, or fax (317) 634-7817.

FOR ADDITIONAL INFORMATION
Contact the American College of Sports Medicine at Tel.: (317) 637-9200, ext. 110, or e-mail: akatzenberger@acsm.org. **For technical questions**, contact the Abstract support desk at Tel.: (217) 398-1792, Fax: (217) 355-0101, or e-mail: support@abstractsonline.com

6   The 10th Measurement and Evaluation Symposium • October 16-18, 2003 • University of Illinois • Urbana-Champaign, Illinois, USA

**AAALF, ACSM, Human Kinetics Publishers and
The University of Illinois at Urbana-Champaign Present:**
The 10th Measurement and Evaluation Symposium:
Measurement Issues and Challenges in Aging Research

10TH MEASUREMENT AND EVALUATION SYMPOSIUM **October 16-18, 2003 • University of Illinois, Urbana-Champaign, Illinois, USA**

REGISTRATION FORM

CONFERENCE DESCRIPTION AND INFORMATION

Established in 1949, the Measurement and Evaluation (M&E) Council is one of the established AAHPERD councils with a long and rich history. This 2003 co-sponsored event will focus on measurement issues and challenges in aging research, including a one-day pre-conference workshop on structural equation modeling and longitudinal study. The conference will take place on the University of Illinois campus in Urbana-Champaign, Illinois. Nearby hotel accommodations have been secured and discounted rates made available.

Instructions:

You may register online, www.acsm.org, or by mail or fax. Type or print your information. Please retain a copy of this form for your records. Completed registration forms should be forwarded to: ACSM, PO Box 663607, Indianapolis, IN 46266. Fax: (317) 634-7817. Registration forms will be accepted through October 4, 2003. After October 4, please plan to register on site.

Registration Information (please print):

Last Name _____   First Name _____

[ ] Male [ ] Female   Date of Birth (optional) |__|__| / __|__| / __|__|

Institution/Affiliation _____

Address _____

City _____   State or Province _____

Zip Code/Postal Code _____   Country _____

Daytime phone _____   Fax _____

E-mail _____   Have you submitted an abstract? _____

Are you planning to attend the Preconference Workshop? |__| Yes |__| No (No additional fee, but space must be reserved)
|__| Indicate if you would like to receive information about sponsorship opportunities for the Measurement and Evaluation Symposium.

Fee Schedule:

	ACSM or AAALF Members	Non-member
[ ] Registration received by September 26, 2003	$275 (U.S. dollars)	$300 (U.S. dollars)
[ ] Registration received after September 26, 2003	$300 (U.S. dollars)	$325 (U.S. dollars)
[ ] Registration for full time students by September 26, 2003	$100 (U.S. dollars)	$110 (U.S. dollars)
[ ] Registration for full time students after September 26, 2003	$125 (U.S. dollars)	$135 (U.S. dollars)

Payment:

Total amount: $_____
[ ] Check enclosed (Make checks payable to ACSM. Federal ID #23-6390952. $25 fee for returned checks)
[ ] Visa or Mastercard
Card Number |__|__|__|__|__|__|__|__|__|__|__|__|__|__|__|__| (list all 13 or 16 numbers)
Expiration Date_____ Signature_____

Please forward to: American College of Sports Medicine, PO Box 663607, Indianapolis, IN 46266, Fax: (317) 634-7817.
For housing, travel, questions and/or additional information, contact ACSM at Tel.: (317) 637-9200, ext. 141, 143, or 135, akrug@acsm.org or ahinkle@acsm.org. **To register via the web, visit www.acsm.org.**

Cancellation Policy: Requests for cancellation must be made in writing on or before September 26, 2003. A $50 fee is retained for processing. No-shows are not considered "cancellations" and receive no refunds.

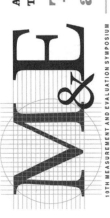

AAALF, ACSM, Human Kinetics Publishers and
The University of Illinois at Urbana-Champaign Present:

The 10th Measurement and Evaluation Symposium

10TH MEASUREMENT AND EVALUATION SYMPOSIUM

Tel.: (317) 637-9200 • FAX: (317) 634-7817 • www.acsm.org
Street Address: 401 W. Michigan St. • Indianapolis, IN 46202-3233 USA
Mailing Address: P.O. Box 663607 • Indianapolis, IN 46266 USA

161

References

Chapter 1

American College of Sports Medicine. (1998). Position Stand. Exercise and physical activity for older adults. *Medicine and Science in Sports and Exercise, 30,* 992-1008.

An, P., Perusse, L., Rankinen, T., Borecki, I.B., Gagnon, J., Leon, A.S., Skinner, J.S., Wilmore, J.H., Bouchard, C., & Rao, D.C. (2003). Familial aggregation of exercise heart rate and blood pressure in response to 20 weeks of endurance training in the HERITAGE Family Study. *International Journal of Sports Medicine, 24,* 57-62.

An, P., Rice, T., Gagnon, J., Leon, A.S., Skinner, J.S., Bouchard, C., Rao, D.C., & Wilmore, J.H. (2000a). Familial aggregation of stroke volume and cardiac output during submaximal exercise: the HERITAGE Family Study. *International Journal of Sports Medicine, 21,* 566-572.

An, P., Rice, T., Perusse, L., Borecki, I., Gagnon, J., Leon, A., Skinner, J., Wilmore, J., Bouchard, C., & Rao, D. (2000b). Complex segregation analysis of blood pressure and heart rate measured before and after a 20-week endurance exercise training program: the HERITAGE Family Study. *American Journal of Hypertension, 13,* 488-497.

Bouchard, C. (1995). Individual differences in the response to regular exercise. *International Journal of Obesity and Related Metabolic Disorders, 19* (Suppl. 4), 5-8.

Bouchard, C., An, P., Rice, T., Skinner, J.S., Wilmore, J.H., Gagnon, J., Perusse, L., Leon, A.S., & Rao, D.C. (1999). Familial aggregation of $\dot{V}O_2$max response to exercise training: results from the HERITAGE Family Study. *Journal of Applied Physiology, 87,* 1003-1008.

Bouchard, C., Daw, E.W., Rice, T., Perusse, L., Gagnon, J., Province, M.A., Leon, A.S., Rao, D.C., Skinner, J.S., & Wilmore, J.H. (1998). Familial resemblance for $\dot{V}O_2$max in the sedentary state: the HERITAGE Family Study. *Medicine and Science in Sports and Exercise, 30,* 252-258.

Bouchard, C., Dionne, F.T., Simoneau, J.A., & Boulay, M.R. (1992). Genetics of aerobic and anaerobic performances. *Exercise and Sport Sciences Reviews, 20,* 27-58.

Bouchard, C., Lesage, R., Lortie, G., Simoneau, J.A., Hamel, P., Boulay, M.R., Perusse, L., Theriault, G., & Leblanc, C. (1986). Aerobic performance in brothers, dizygotic and monozygotic twins. *Medicine and Science in Sports and Exercise, 18,* 639-646.

Bouchard, C., & Rankinen, T. (2001). Individual differences in response to regular physical activity. *Medicine and Science in Sports and Exercise, 33* (Suppl.), S446-S451.

Bronikowski, A.M., Carter, P.A., Morgan, T.J., Garland Jr., T., Ung, N., Pugh, T.D., Weindruch, R., & Prolla, T.A. (2003). Lifelong voluntary exercise in the mouse prevents age-related alterations in gene expression in the heart. *Physiological Genomics, 12,* 129-138.

Chagnon, Y.C., Rankinen, T., Snyder, E.E., Weisnagel, S.J., Perusse, L., & Bouchard, C. (2003). The Human Obesity Gene Map: the 2002 Update. *Obesity Research, 11,* 313-367.

Fagard, R., Bielen, E., & Amery, A. (1991). Heritability of aerobic power and anaerobic energy generation during exercise. *Journal of Applied Physiology, 70,* 357-362.

Laitinen, P.J., Brown, K.M., Piippo, K., Swan, H., Devaney, J.M., Brahmbhatt, B., Donarum, E.A., Marino, M., Tiso, N., Viitasalo, M., Toivonen, L., Stephan, D.A., & Kontula, K. (2001). Mutations of the cardiac ryanodine receptor (RyR2) gene in familial polymorphic ventricular tachycardia. *Circulation, 103,* 485-490.

Leon, A.S., Rice, T., Mandel, S., Despres, J.P., Bergeron, J., Gagnon, J., Rao, D.C., Skinner, J.S., Wilmore, J.H., & Bouchard, C. (2000). Blood lipid response to 20 weeks of supervised exercise in a large biracial population: the HERITAGE Family Study. *Metabolism, 49,* 513-520.

Lutz, W., Sanderson, W., & Scherbov, S. (2001). The end of world population growth. *Nature, 412,* 543-545.

Perls, T.T., Wilmoth, J., Levenson, R., Drinkwater, M., Cohen, M., Bogan, H., Joyce, E., Brewster, S., Kunkel, L., & Puca, A. (2002). Life-long sustained mortality advantage of siblings of centenarians. *Proceedings of the National Academy of Sciences USA, 99,* 8442-8447.

Perusse, L., Gagnon, J., Province, M.A., Rao, D.C., Wilmore, J.H., Leon, A.S., Bouchard, C., & Skinner, J.S. (2001). Familial aggregation of submaximal aerobic performance in the HERITAGE Family Study. *Medicine and Science in Sports and Exercise, 33,* 597-604.

Perusse, L., Rankinen, T., Rauramaa, R., Rivera, M.A., Wolfarth, B., & Bouchard, C. (2003). The Human Gene Map for performance and health-related fitness phenotypes: the 2002 Update. *Medicine and Science in Sports and Exercise, 35* (8), 1248-1264.

Perusse, L., Rice, T., Province, M.A., Gagnon, J., Leon, A.S., Skinner, J.S., Wilmore, J.H., Rao, D.C., & Bouchard, C. (2000). Familial aggregation of amount and distribution of subcutaneous fat and their responses to exercise training in the HERITAGE family study. *Obesity Research, 8,* 140-150.

Puca, A.A., Daly, M.J., Brewster, S.J., Matise, T.C., Barrett, J., Shea-Drinkwater, M., Kang, S., Joyce, E., Nicoli, J., Benson, E., Kunkel, L.M., & Perls, T. (2001). A genome-wide scan for linkage to human exceptional longevity identifies a locus on chromosome 4. *Proceedings of the National Academy of Sciences USA, 98,* 10505-10508.

Rankinen, T., Gagnon, J., Perusse, L., Chagnon, Y., Rice, T., Leon, A., Skinner, J., Wilmore, J., Rao, D., & Bouchard, C. (2000). AGT M235T and ACE ID polymorphisms and exercise blood pressure in the HERITAGE Family Study. *American Journal of Physiology: Heart and Circulatory Physiology, 279,* H368-374.

Rao, D.C., & Province, M.A. (Eds.). (2001). *Genetic Dissection of Complex Traits.* San Diego: Academic Press.

Rauramaa, R., Kuhanen, R., Lakka, T.A., Vaisanen, S.B., Halonen, P., Alen, M., Rankinen, T., & Bouchard, C. (2002). Physical exercise and blood pressure with reference to the angiotensinogen M235T polymorphism. *Physiological Genomics, 10,* 71-77.

Rice, T., An, P., Gagnon, J., Leon, A., Skinner, J., Wilmore, J., Bouchard, C., & Rao, D. (2002a). Heritability of HR and BP response to exercise training in the HERITAGE Family Study. *Medicine and Science in Sports and Exercise, 34,* 972-979.

Rice, T., Despres, J.P., Perusse, L., Hong, Y., Province, M.A., Bergeron, J., Gagnon, J., Leon, A.S., Skinner, J.S., Wilmore, J.H., Bouchard, C., & Rao, D.C. (2002b). Familial aggregation of blood lipid response to exercise training in the health, risk factors, exercise training, and genetics (HERITAGE) Family Study. *Circulation, 105,* 1904-1908.

Rice, T., Hong, Y., Perusse, L., Despres, J.P., Gagnon, J., Leon, A.S., Skinner, J.S., Wilmore, J.H., Bouchard, C., & Rao, D.C. (1999). Total body fat and abdominal visceral fat response to exercise training in the HERITAGE Family Study: evidence for major locus but no multifactorial effects. *Metabolism, 48,* 1278-1286.

Schachter, F., Faure-Delanef, L., Guenot, F., Rouger, H., Froguel, P., Lesueur-Ginot, L., & Cohen, D. (1994). Genetic associations with human longevity at the APOE and ACE loci. *Nature Genetics, 6,* 29-32.

Skinner, J.S., Wilmore, K.M., Krasnoff, J.B., Jaskolski, A., Jaskolska, A., Gagnon, J., Province, M.A., Leon, A.S., Rao, D.C., Wilmore, J.H., & Bouchard, C. (2000). Adaptation to a standardized training program and changes in fitness in a large, heterogeneous population: the HERITAGE Family Study. *Medicine and Science in Sports and Exercise, 32,* 157-161.

Sundet, J.M., Magnus, P., & Tambs, K. (1994). The heritability of maximal aerobic power: a study of Norwegian twins. *Scandinavian Journal of Medicine and Science in Sports, 4,* 181-185.

Swan, H., Piippo, K., Viitasalo, M., Heikkila, P., Paavonen, T., Kainulainen, K., Kere, J., Keto, P., Kontula, K., & Toivonen, L. (1999). Arrhythmic disorder mapped to chromosome 1q42-q43 causes malignant polymorphic ventricular tachycardia in structurally normal hearts. *Journal of the American College of Cardiology, 34,* 2035-2042.

Wang, Q., Curran, M.E., Splawski, I., Burn, T.C., Millholland, J.M., VanRaay, T.J., Shen, J., Timothy, K.W., Vincent, G.M., de Jager, T., Schwartz, P.J., Toubin, J.A., Moss, A.J., Atkinson, D.L., Landes, G.M., Connors, T.D., & Keating, M.T. (1996). Positional cloning of a novel potassium channel gene: KVLQT1 mutations cause cardiac arrhythmias. *Nature Genetics, 12,* 17-23.

Welle, S., Bhatt, K., & Thornton, C.A. (2000). High-abundance mRNAs in human muscle: comparison between young and old. *Journal of Applied Physiology, 89,* 297-304.

Wilmore, J.H., Stanforth, P.R., Gagnon, J., Rice, T., Mandel, S., Leon, A.S., Rao, D.C., Skinner, J.S., & Bouchard, C. (2001). Heart rate and blood pressure changes with endurance training: the HERITAGE Family Study. *Medicine and Science in Sports and Exercise, 33,* 107-116.

Chapter 2

American College of Sports Medicine. (1998). Position Stand: the recommended quantity and quality of exercise for developing and maintaining cardiorespiratory and muscular fitness and flexibility in healthy adults. *Medicine and Science in Sports and Exercise, 30,* 975-991.

Baker, K., Nelson, M., Felson, D., Layne, J., Sarno, R., & Roubenoff, R. (2001). The efficacy of home based progressive strength training in older adults with knee osteoarthritis: a randomized controlled trial. *Journal of Rheumatology, 28,* 1655-1665.

Baumgartner, R., Koehler, K., Gallagher, D., Romero, L., Heymsfield, S.B., Ross, R.R., Garry, P.J., & Lindeman, R.D. (1998). Epidemiology of sarcopenia among the elderly in New Mexico. *American Journal of Epidemiology, 147,* 755-763.

Beniamini, Y., Rubenstein, J., Faigenbaum, A., Lichtenstein, A., & Crim, M. (1999). High-intensity strength training of patients enrolled in an outpatient cardiac rehabilitation program. *Journal of Cardiopulmonary Rehabilitation, 19,* 8-17.

Buchner, D.M., Beresford, S.A.A., Larson, A.B., LaCroix, A.Z., & Wagner, E.H. (1992). Effects of physical activity on health status in older adults II: intervention studies. *Annual Review of Public Health, 13,* 469-488.

Buchner, D., Cress, M., de Lateur, B., Esselman, P.C., Margherita, A.J., Price, R., & Wagner, E.H. (1997). The effect of strength and endurance training on gait, balance, fall risk, and health services use in community-living older adults. *Journal of Gerontology, 52* (4), 218-224.

Campbell, A., Robertson, C., Gardner, M., Norton, R., & Buchner, D. (1999). Falls prevention over 2 years: a randomized controlled trial in women 80 years and older. *Age and Ageing, 28,* 513-528.

Campbell, A., Robertson, M., Gardner, M., Norton, R., Tilyard, M., & Buchner, D. (1997). Randomized controlled trial of a general practice programme of home based exercise to prevent falls in elderly women. *British Medical Journal, 315,* 1065-1069.

Carlson, K.A., & Harshman, L.G. (1999). Extended longevity lines of Drosophila melanogaster: characterization of oocyte stages and ovariole numbers as a function of age and diet. *Journal of Gerontology, 54,* 432-440.

Castaneda, C., Layne, J., Munoz-Orians, L., Gordon, P.L., Walsmith, J., Foldvari, M., Roubenoff, R., Tucker, K.L., & Nelson, M.E. (2002). A randomized control trial of progressive resistance exercise training in older adults with type 2 diabetes. *Diabetes Care, 25,* 2335-2341.

Chakravarthy, M., Joyner, M., & Booth, F. (2002). An obligation for primary care physicians to prescribe physical activity to sedentary patients to reduce the risk of chronic health conditions. *Mayo Clinic Proceedings, 77,* 165-173.

Cussler, E., Lohman, T., Going, S., Houtkooper, L.B., Metcalfe, L.L., Flint-Wagner, H.G., Harris, R.B., & Teixeira, P.J. (2003). Weight lifted in strength training predicts bone change in postmenopausal women. *Medicine and Science in Sports and Exercise, 35,* 10-17.

Dunstan, D., Daly, R., Owen, N., Jolley, D., De Courten, M., Shaw, J., & Zimmet, P. (2002). High-intensity resistance training improves glycemic control in older patients with type 2 diabetes. *Diabetes Care, 25,* 1729-1736.

Ettinger, W., Burns, R., Messier, S., Applegate, W., Rejeski, W.J., Morgan, T., Shumaker, S., Berry, M.J., O'Toole, M., Monu, J., & Craven, T. (1997). A randomized trial comparing aerobic exercise and resistance exercise with a health education program in older adults with knee osteoarthritis: the fitness arthritis and seniors trial (FAST). *Journal of the American Medical Association, 277* (1), 25-31.

Ferrucci, L., Izmirlian, G., Leveille, S., Phillips, C.L., Corti, M., Brock, D.B., & Guralnik, J.M. (1999). Smoking, physical activity, and active life expectancy. *American Journal of Epidemiology, 149* (7), 645-653.

Fiatarone, M., Marks, E., Ryan, N., Meredith, C., Lipsitz, L., & Evans, W. (1990). High-intensity strength training in nonagenarians. Effects on skeletal muscle. *Journal of the American Medical Association, 263,* 3029-3034.

Fiatarone, M., O'Neill, E., Ryan, N., Clements, K.M., Solares, G.R., Nelson, M.E., Roberts, S.B., Kehayias, J.J., Lipsitz, L.A., & Evans, W.J. (1994). Exercise training and nutritional supplementation for physical frailty in very elderly people. *New England Journal of Medicine, 330* (25), 1769-1775.

Fielding, R., LeBrasseur, N., Cuoco, A., Bean, J., Mizer, K., & Singh, M. (2002). High-velocity resistance training increases skeletal muscle peak power in older women. *Journal of the American Geriatrics Society, 50,* 655-662.

Foldvari, M., Clark, M., Laviolette, L., Bernstein, M.A., Kaliton, D., Castaneda, C., Pu, C.T., Hausdorff, J.M., Fielding, R.A., & Singh, M.A. (2000). Association of muscle power with functional status in community-dwelling elderly women. *Journal of Gerontology, 55A,* M192-M199.

Frontera, W., Merideth, C., O'Reilly, K., Knuttgen, H., & Evans, W. (1988). Strength conditioning in older men: skeletal muscle hypertrophy and improved function. *Journal of Applied Physiology, 64* (3), 1038-1044.

Janssen, I., Heymsfield, S., Wang, Z., & Ross, R. (2000). Skeletal muscle mass and distribution in 468 men and women aged 18-88 yr. *Journal of Applied Physiology, 89,* 81-88.

Jette, A., Lachman, M., Giorgetti, M., Assmann, S.F., Harris, B.A., Levenson, C., Wernick, M., & Krebs, D. (1999). Exercise—it's never too late: the strong for life program. *American Journal of Public Health, 89,* 66-72.

Kohrt, W., Ehsani, A., & Birge, S. (1997). Effects of exercise involving predominantly either joint-reaction or ground-reaction forces on bone mineral density in older women. *Journal of Bone and Mineral Research, 12,* 1253-1261.

LaCroix, A.Z., Guralnik, J.M., Berkman, L.F., Wallace, R.B., & Satterfield, S. (1993). Maintaining mobility in late life. II. Smoking, alcohol consumption, physical activity, and body mass index. *American Journal of Epidemiology, 137,* 858-869.

Leveille, S.G., Guralnik, J.M., Ferrucci, L., & Langlois, J.A. (1999). Aging successfully until death in old age: opportunities for increasing active life expectancy. *American Journal of Epidemiology, 149* (7), 654-664.

Manton, K., & Gu, X. (2001). Changes in the prevalence of chronic disability in the United States black and nonblack population above age 65 from 1982 to 1999. *Proceedings of the National Academy of Sciences USA, 98,* 6354-6359.

McCartney, N., Hicks, A.L., Martin, J., & Webber, C.E. (1995). Long-term resistance training in the elderly: effects on dynamic strength, exercise capacity, muscle and bone. *Journal of Gerontology, 50,* 97-104.

Messier, S., Royer, T., Craven, T., O'Toole, M., Burns, R., & Ettinger, W. (2000). Long-term exercise and its effect on balance in older, osteoarthritic adults: results from the Fitness, Arthritis, and Seniors Trial (FAST). *Journal of the American Geriatrics Society, 48,* 131-138.

Nelson, M., Fiatarone, M., Morganti, C., Trice, I., Greenberg, R., & Evans, W. (1994). Effects of high-intensity strength training on multiple risk factors for osteoporotic fractures. *Journal of the American Medical Association, 272,* 1909-1914.

Nelson, M., Layne, J., Bernstein, M., Nuernberger, A., Castaneda, C., Kaliton, D., Hausdorff, J., Judge, J.O., Buchner, D.M., Roubenoff, R., & Singh, M.A. (2004). The effects of multidimensional home-based exercise on functional performance in the elderly. *Journals of Gerontology Series A: Biological Sciences and Medical Sciences, 59* (2), 54-60.

Rosenberg, I. (1989). Summary comments. *American Journal of Clinical Nutrition, 50,* 1231-1233.

Roubenoff, R. (2000). Sarcopenia and its implications for the elderly. *European Journal of Clinical Nutrition, 54* (Suppl.), 40-47.

Roubenoff, R., & Hughes, V. (2000). Sarcopenia: current concepts. *Journal of Gerontology, 55A,* M716-M724.

Rubenstein, L., Josephson, K., Trueblood, P., Loy, S., Harker, J.O., Pietruszka, F.M., & Robbins, A.S. (2000). Effects of a group exercise program on strength, mobility, and falls among fall-prone elderly men. *Journal of Gerontology, 55,* 317-321.

Sevick, M., Bradham, D., Meunder, M., Chen, G.J., Enarson, C., Dailey, M., & Ettinger, W.H. Jr. (2000). Cost-effectiveness of aerobic and resistance training in seniors with knee osteoarthritis. *Medicine and Science in Sports and Exercise, 32,* 1534-1540.

Singh, N., Clements, K., & Fiatarone, M. (1997a). A randomized controlled trial of progressive resistance training in depressed elders. *Journal of Gerontology, 52,* 27-35.

Singh, N., Clements, K., & Fiatarone, M. (1997b). Sleep, sleep deprivation, and daytime activities: a randomized controlled trial of the effect of exercise on sleep. *Sleep, 20,* 95-101.

Skelton, D., Young, A., Greig, C., & Malbut, K. (1995). Effects of resistance training on strength, power, and selected functional abilities of women aged 75 and older. *Journal of the American Geriatrics Society, 43,* 1081-1087.

Strawbridge, W.J., Cohen, R.D., Shema, S.J., & Kaplan, G.A. (1996). Successful aging: predictors and associated activities. *American Journal of Epidemiology, 144,* 135-141.

Taaffe, D., Duret, C., Wheeler, S., & Robert, M. (1999). Once-weekly resistance exercise improves muscle strength and neuromuscular performance in older adults. *Journal of the American Geriatrics Society, 47,* 1208-1214.

Thomas, K., Muir, K., Doherty, M., Jones, A., O'Reilly, S., & Bassey, E. (2002). Home based exercise programme for knee pain and knee osteoarthritis: randomised controlled trial. *British Medical Journal,* October 5, 325(7367): 752.

U.S. Department of Health and Human Services. (2000). *Healthy People 2010: Understanding and Improving Health.* Washington, DC: U.S. Government Printing Office.

Wu, S.C., Leu, S., & Li, C. (1999). Incidence of and predictors for chronic disability in activities of daily living among older people in Taiwan. *Journal of the American Geriatrics Society, 47* (9), 1082-1086.

Chapter 3

Bahr, R., & Sejersted, O.M. (1991). Effect of intensity of exercise on excess postexercise O_2 consumption. *Metabolism, 40,* 836-841.

Bijnen, F.C.H., Caspersen, C.J., & Mosterd, W.L. (1995). Physical inactivity as a risk factor for coronary heart disease: a WHO and International Society and Federation of Cardiology position statement. *Bulletin of the World Health Organisation, 72,* 1-4.

Blair, S.N., Cheng, Y., & Holder, J.S. (2001). Is physical activity or physical fitness more important in defining health benefits. *Medicine and Science in Sports and Exercise, 33* (Suppl. 6), S379-S399.

Blair, S.N., & Connelly, J.C. (1996). How much physical activity should we do? The case for moderate amounts and intensities of physical activity. *Research Quarterly, 67,* 193-205.

Blair, S.N., Kohl, H.W., Barlow, C.E., Paffenbarger, R.S., Gibbons, L.W., & Macera, C.A. (1995). Changes in physical fitness and all-cause mortality: a prospective study of healthy and unhealthy men. *Journal of the American Medical Association, 273,* 1093-1098.

Blair, S.N., Kohl, H.W., Paffenbarger, R.S., Clark, D.G., Cooper, K.H., & Gibbons, L.W. (1989). Physical fitness and all-cause mortality: a prospective study of healthy men and women. *Journal of the American Medical Association, 262,* 2395-2401.

Bouchard, C., Shephard, R.J., & Stephens, T. (1994). *Physical Activity, Fitness, and Health.* Champaign, IL: Human Kinetics.

Bouchard, C., Shephard, R.J., Stephens, T., Sutton, J., & McPherson, B. (1990). *Exercise, Fitness, and Health*. Champaign, IL: Human Kinetics.

Braith, R.W., Pollock, M.L., Lowenthal, D.T., Graves, J.E., & Limacher, M.C. (1994). Moderate- and high-intensity exercise lowers blood pressure in normotensive subjects 60 to 79 years of age. *American Journal of Cardiology, 73*, 1124-1128.

Braun, B., Zimmerman, M.B., & Kretchmer, N. (1995). Effects of exercise intensity on insulin sensitivity in women with non-insulin-dependent diabetes mellitus. *Journal of Applied Physiology, 78*, 300-306.

Broeder, C.E., Brenner, M., Hofman, Z., Paijmans, I.J., Thomas, E.L., & Wilmore, J.H. (1991). The metabolic consequences of low and moderate intensity exercise with or without feeding in lean and borderline obese males. *International Journal of Obesity, 15*, 95-104.

Chad, K.E., & Quigley, B.M. (1991). Exercise intensity: effect on postexercise O_2 uptake in trained and untrained women. *Journal of Applied Physiology, 70*, 1713-1719.

Duncan, J.J., Gordon, N.F., & Scott, C.B. (1991). Women walking for health and fitness. How much is enough? *Journal of the American Medical Association, 266*, 3295-3299.

Dunn, A.L., Andersen, R.E., & Jakicic, J.M. (1998). Lifestyle physical activity interventions. History, short- and long-term effects, and recommendations. *American Journal of Preventive Medicine, 15*, 398-412.

Folsom, A.R., Prineas, R.J., Kaye, S.A., & Munger, R.G. (1990). Incidence of hypertension and stroke in relation to body fat distribution and other risk factors in older women. *Stroke, 21*, 701-706.

Friedlander, A.L., Casazza, G.A., Horning, M.A., Buddinger, T.F., & Brooks, G.A. (1998). Effects of exercise intensity and training on lipid metabolism in young women. *American Journal of Physiology, 275*, E853-E863.

Goben, K.W., Sforzo, G.A., & Frye, P.A. (1992). Exercise intensity and the thermic effect of food. *International Journal of Sport Nutrition, 2*, 87-95.

Haapanen, N., Miilunpalo, S., Vuori, I., Oja, P., & Pasanen, M. (1998). Association of leisure time physical activity with the risk of coronary heart disease, hypertension and diabetes in middle aged men and women. *International Journal of Epidemiology, 26*, 739-747.

Hagberg, J.M., Montain, S.J., Martin, W.H., & Ehsani, M.A. (1989). Effect of exercise training in 60 to 69-year-old persons with essential hypertension. *American Journal of Cardiology, 64*, 348-353.

Health Canada. (1999). *Handbook for Canada's Physical Activity Guide to Healthy Active Living*. Ottawa, ON: Health Canada/Canadian Society of Exercise Physiology.

Herman, B., Schmitz, P.I.M., Leyten, A.C.M., Van Luijk, J.H., Frenken, C.W.G.M., Op de Coul, A.A., & Schulte, B.P. (1983). Multivariate logistic analysis of risk factors for stroke in Tilburg, the Netherlands. *American Journal of Epidemiology, 118*, 514-525.

Howley, E.T. (2001). Type of activity: resistance, aerobic, anaerobic and leisure-time versus occupational physical activity. *Medicine and Science in Sports and Exercise, 33* (Suppl. 6), S364-S369.

Kesaniemi, Y.A., Danforth, E., Jensen, M.D., Kopelman, P.G., Lefebvre, P., & Reeder, B.A. (2001). Dose-response issues concerning physical activity and health: an evidence-based symposium. *Medicine and Science in Sports and Exercise, 33* (Suppl. 6), S351-S358.

Kiely, D.K., Wolf, P.A., Cupples, L.A., Beiser, A.S., & Kannel, W.B. (1994). Physical activity and stroke risk: the Framingham study. *American Journal of Epidemiology, 140*, 608-620.

Kingwell, B.A., & Jennings, G. (1993). Effects of walking and other exercise programs upon blood pressure in normal subjects. *Medical Journal of Australia, 158,* 234-238.

Lee, I.-M., Manson, J.E., Ajani, U., Paffenbarger, R.S., Hennekens, C.H., & Buring, J.E. (1997). Physical activity and risk of colon cancer: the Physicians' Health Study (United States). *Cancer Causation and Control, 8,* 568-574.

Lee, I.-M., & Paffenbarger, R.S. (1996). How much physical activity is optimal for health? Methodological considerations. *Research Quarterly, 67,* 206-208.

Lee, I.-M., & Paffenbarger, R. (1997). Is vigorous physical activity necessary to reduce the risk of cardiovascular disease? In A.S. Leon (Ed.), *Physical Activity and Cardiovascular Health* (pp. 67-75). Champaign, IL: Human Kinetics.

Linsted, K.D., Tonstad, S., & Kuzma, J.W. (1991). Self-report of physical activity and patterns of mortality in Seventh-day Adventist men. *Journal of Clinical Epidemiology, 44,* 355-364.

Manson, J.E., Hu, F.B., Rich-Edwards, J.W., Colditz, G.A., Stampfer, M.J., Willett, W.C., Speizer, F.E., & Hennekens, C.H. (1999). A prospective study of walking as compared with vigorous exercise in the prevention of coronary heart disease in women. *New England Journal of Medicine, 341,* 650-658.

Matsusaki, M., Ikeda, M., Tashiro, E., Koga, M., Miura, S., Ideishi, M., Tanaka, H., Shindo, M., & Arakawa, K. (1992). Influence of workload on the antihypertensive effect of exercise. *Clinical and Experimental Pharmacology and Physiology, 19,* 471-479.

Menotti, A., & Seccareccia, F. (1985). Physical activity at work and job as risk factors for fatal coronary heart disease and other causes of death. *Journal of Epidemiology and Community Health, 39,* 325-329.

Mittelman, M.A., Maclure, M., Tofler, G.H., Sherwood, J.B., Goldberg, R.J., & Muller, J.E. (1993). Triggering of acute myocardial infarction by heavy physical exertion: protection against triggering by regular exertion. *New England Journal of Medicine, 329,* 1677-1683.

Paffenbarger, R.S., Hyde, R.T., Wing, A.L., & Hsieh, C.C. (1986). Physical activity, all-cause mortality, and longevity of college alumni. *New England Journal of Medicine, 314,* 605-613.

Paffenbarger, R.S., Wing, A.L., Hyde, R.T., & Jung, D.L. (1983). Physical activity and incidence of hypertension in college alumni. *American Journal of Epidemiology, 117,* 245-257.

Pruitt, L.A., Taaffe, D.R., & Marcus, R. (1995). Effects of a one-year high-intensity versus low-intensity resistance training program on bone mineral density in older women. *Journal of Bone and Mineral Research, 10,* 1788-1795.

Rogers, M.W., Probst, M.M., Gruber, J.J., Berger, R., & Boone, J.B. (1996). Differential effects of exercise training intensity on blood pressure and cardiovascular responses to stress in borderline hypertensive humans. *Journal of Hypertension, 14,* 1369-1375.

Roman, O., Camuzzi, A.L., Villalon, E., & Klenner, C. (1981). Physical training program in arterial hypertension. A long-term prospective follow-up. *Cardiology, 67,* 230-243.

Rosengren, A., & Wilhelmsen, L. (1997). Physical activity protects against coronary death and deaths from all causes in middle-aged men. Evidence from a 20-year follow-up of the primary prevention study in Göteborg. *Annals of Epidemiology, 7,* 69-75.

Sacco, R.L., Gan, R., Boden-Albala, B., Lin, I.F., Kargman, D.E., Hauser, W.A., Shea, S., & Paik, M.C. (1998). Leisure-time physical activity and ischemic stroke risk: the Northern Manhattan Stroke Study. *Stroke, 29,* 380-387.

Sedlock, D.A. (1991). Effect of exercise intensity on postexercise energy expenditure in women. *British Journal of Sports Medicine, 25,* 38-40.

Segal, K.R., Chun, A., Coronel, P., & Valdez, V. (1992). Effects of exercise mode and intensity on postprandial thermogenesis in lean and obese men. *Journal of Applied Physiology, 72,* 1754-1763.

Shephard, R.J. (1965). The development of cardiorespiratory fitness. *Medical Services Journal, Canada, 21,* 533-544.

Shephard, R.J. (1968). Intensity, duration and frequency of exercise as determinants of the response to a training regime. *Internationale Zeitschrift für angewandte Physiologie, 26,* 272-278.

Shephard, R.J. (1994a). *Aerobic Fitness and Health.* Champaign, IL: Human Kinetics.

Shephard, R.J. (1994b). Determinants of exercise in people aged 65 years and older. In R. Dishman (Ed.), *Advances in Exercise Adherence* (pp. 343-360). Champaign, IL: Human Kinetics.

Shephard, R.J. (1996). Habitual physical activity and the quality of life. *Quest, 48,* 354-365.

Shephard, R.J. (1997a). *Aging, Physical Activity, and Health.* Champaign, IL: Human Kinetics.

Shephard, R.J. (1997b). *Physical Activity, Training and the Immune Response.* Carmel, IN: Cooper.

Shephard, R.J. (1997c). What is the optimal type of physical activity to enhance health? *British Journal of Sports Medicine, 31,* 277-284.

Shephard, R.J. (1999). What is the optimal type of physical activity to enhance health? In D. MacAuley (Ed.), *Benefits and Hazards of Exercise* (pp. 1-24). London, UK: BMJ Books.

Shephard, R.J. (2001). Relative vs. absolute intensity of exercise in a dose-response context. *Medicine and Science in Sports and Exercise, 33* (Suppl. 6), S400-S418.

Shephard, R.J. (2002). Whistler 2001. A Health Canada/CDC conference on "Communicating Physical Activity and Health Messages: Science into Practice." *American Journal of Preventive Medicine, 23,* 221-224.

Shephard, R.J. (2003). Regression to the mean: a threat to exercise science? *Sports Medicine, 33,* 575-584.

Shephard, R.J., & Bouchard, C. (1995). Relationship between perceptions of physical activity and health-related fitness. *Journal of Sports Medicine and Physical Fitness, 35,* 149-158.

Shephard, R.J., & Bouchard, C. (1996). Associations between health behaviours and health related fitness. *British Journal of Sports Medicine, 30,* 94-101.

Shephard, R.J., & Futcher, R. (1997). Physical activity and cancer: how may protection be maximized? *Critical Reviews in Oncogenesis, 8,* 219-272.

Shephard, R.J., Kavanagh, T., & Mertens, D.J. (1995). The personal benefits of Masters athletic competition. *British Journal of Sports Medicine, 29,* 35-40.

Short, K.R., Wiest, J.M., & Sedlock, D.A. (1996). The effect of upper body exercise intensity and duration on post-exercise oxygen consumption. *International Journal of Sports Medicine, 17,* 559-563.

Smith, J., & McNaughton, L. (1993). The effects of exercise intensity on excess postexercise oxygen consumption and energy expenditure in moderately trained men and women. *European Journal of Applied Physiology, 67,* 420-425.

Tashiro, E., Miura, S., Koga, M., Sasaguri, M.I.M., Ikeda, M., Tanaka, H., Shindo, M., & Arakawa, K. (1993). Crossover comparison between the depressor effects of low and high work rate exercise in mild hypertension. *Clinical and Experimental Pharmacology and Physiology, 20,* 689-696.

Treuth, M.S., Hunter, G.R., & Williams, M. (1996). Effects of exercise intensity on 24h energy expenditure and substrate oxidation. *Medicine and Science in Sports and Exercise, 28,* 1138-1143.

Wannamethee, G., & Shaper, A.G. (1992). Physical activity and stroke in British middle aged men. *British Medical Journal, 304,* 597-601.

Welle, S., Thornton, C., Jozefowicz, R., & Statt, M. (1993). Myofibrillar protein synthesis in young and old men. *American Journal of Physiology, 264,* E693-E698.

Williams, P.T. (1998). Relationships of heart disease risk factors to exercise quantity and intensity. *Archives of Internal Medicine, 158,* 237-245.

Chapter 4

Amrhein, P.C., Goggin, N.L., & Stelmach, G.E. (1991). Age differences in the maintenance and restructuring of movement preparation. *Psychology and Aging, 6* (3), 451-466.

Amrhein, P.C., & Theios, J. (1993). The time it takes elderly and young individuals to draw pictures and write words. *Psychology of Aging, 8* (2), 197-206.

Bellgrove, M.A., Phillips, J.G., Bradshaw, J.L., & Galluci, R.M. (1998). Response (re-) programming in aging: a kinematic analysis. *Journal of Gerontology, 53A* (3), M222-M227.

Bennett, K.M.B., & Castiello, U. (1994). Reach to grasp: changes with age. *Journal of Gerontology, 49B* (1), P1-P7.

Birren, J.E. (1974). Translations in gerontology—from lab to life. Psychophysiology and speed of response. *American Psychology, 29* (11), 808-815.

Brown, S.H. (1996). Control of simple arm movements in the elderly. In A.M. Fernandez & N. Teasdale (Eds.), *Changes in Sensory Motor Behavior in Aging 114.* Amsterdam: Elsevier Science BV (North-Holland).

Campbell, M.J., McComas, A.J., & Petito, F. (1973). Physiological changes in ageing muscles. *Journal of Neurology, Neurosurgery, and Psychiatry, 36* (2), 174-182.

Carnahan, H., Vandervoort, A.A., & Swanson, L.R. (1998). The influence of aging and target motion on the control of prehension. *Experimental Aging Research, 24,* 289-306.

Cerella, J. (1985). Information processing rates in the elderly. *Psychological Bulletin, 98* (1), 67-83.

Cerella, J., Poon, L.W., & Williams, D.M. (1980). Age and the complexity hypothesis. In L. Poon (Ed.), *Aging in the 1980's: Psychological Issues.* Washington, DC: American Psychological Association.

Chaput, S., & Proteau, L. (1996). Modification with aging in the role played by vision and proprioception for movement control. *Experimental Aging Research, 22,* 1-21.

Churchill, J.D., Galvez, R., Colcombe, S., Swain, R.A., Kramer, A.F., & Greenough, W.T. (2002). Exercise, experience and the aging brain. *Neurobiology of Aging, 23* (5), 941-955.

Clamann, H.P. (1993). Motor unit recruitment and the gradation of muscle force. *Physical Therapy, 73* (12), 830-843.

Contreras-Vidal, J.L., Tuelings, H.L., & Stelmach, G.E. (1998). Elderly subjects are impaired in spatial coordination in fine motor control. *Acta Psychologica, 100* (1-2), 25-35.

Cooke, J.D., Brown, S.H., & Cunningham, D.A. (1989). Kinematics of arm movements in elderly humans. *Neurobiology of Aging, 10,* 159-165.

Czaja, S.J., & Sharit, J. (1998). Ability-performance relationships as a function of age and task experience for a data entry task. *Journal of Experimental Psychology: Applied, 4* (4), 332-351.

Darling, W.G., Cooke, J.D., & Brown, S.H. (1989). Control of simple arm movements in elderly humans. *Neurobiology of Aging, 10,* 149-157.

Davies, R.J., & White, M.J. (1983). The contractile properties of elderly human triceps surae. *Gerontology, 29,* 19-23.

Dixon, R.A., Kurzman, D., & Friesen, I.C. (1993). Handwriting performance in younger and older adults: age, familiarity, and practice effects. *Psychology and Aging, 8* (3), 360-370.

Doherty, T.J., Vandervoort, A.A., & Brown, W.F. (1993). Effects of ageing on the motor unit: a brief review. *Canadian Journal of Applied Physiology, 18* (4), 331-358.

Dounskaia, N.V., Ketcham, C.J., & Stelmach, G.E. (2002a). Commonalities and differences in control of various drawing movements. *Experimental Brain Research, 146,* 11-25.

Dounskaia, N.V., Ketcham, C.J., & Stelmach, G.E. (2002b). Influence of biomechanical constraints on horizontal arm movements. *Motor Control, 6,* 366-387.

Erim, Z., Beg, M.F., Burke, D.T., & De Luca, C.J. (1999). Effects of aging on motor-unit control properties. *Journal of Neurophysiology, 82,* 2081-2091.

Etnier, J.L., & Landers, D.M. (1998). Motor performance and motor learning as a function of age and fitness. *Research Quarterly for Exercise and Sport, 69* (2), 136-146.

Fozard, J.L., Vercruyssen, M., Reynolds, S.L., Hancock, P.A., & Quilter, R.E. (1994). Age differences and changes in reaction time: the Baltimore Longitudinal Study of Aging. *Journal of Gerontology, B 49* (4), P179-P189.

Galganski, M.E., Fuglevand, A.J., & Enoka, R.M. (1993). Reduced control of motor output in a human hand muscle of elderly subjects during submaximal contractions. *Journal of Neurophysiology, 69* (6), 2108-2115.

Goggin, N.L., & Meeuwsen, H.J. (1992). Age-related differences in the control of spatial aiming movements. *Research Quarterly for Exercise and Sport, 63* (4), 356-372.

Goggin, N.L., & Stelmach, G.E. (1990). Age-related differences in a kinematic analysis of precued movements. *Canadian Journal on Aging, 9,* 371-385.

Gottlob, L.R., & Madden, D.J. (1999). Age differences in the strategic allocation of visual attention. *Journal of Gerontology, 54B* (3), P165-P172.

Greene, L.S., & Williams, H.G. (1996). Aging and coordination from the dynamic pattern perspective. In A.M. Fernandez & N. Teasdale (Eds.), *Changes in Sensory Motor Behavior in Aging.* Amsterdam: Elsevier Science BV (North-Holland).

Greenwood, M., Meeuwsen, H., & French, R. (1993). Effects of cognitive learning strategies, verbal reinforcement, and gender on the performance of closed motor skills in older adults. *Activities, Adaptation and Aging, 17* (3), 39-53.

Gutman, S.R., Latash, M.L., Almeida, G.L., & Gottlieb, G.L. (1993). Kinematic description of variability of fast movements: analytical and experimental approaches. *Biological Cybernetics, 69,* 485-492.

Haaland, K.Y., Harrington, D.L., & Grice, J.W. (1993). Effects of aging on planning and implementing arm movements. *Psychology and Aging, 8* (4), 617-632.

Hakkinen, K., Kraemer, W.J., Kallinen, M., Linnamo, V., Pastinen, U.M., & Newton, R.U. (1996). Bilateral and unilateral neuromuscular function and muscle cross-sectional area in middle-aged and elderly men and women. *Journal of Gerontology, 51A* (1), B21-B29.

Harrington, D.L., & Haaland, K.Y. (1992). Skill learning in the elderly: diminished implicit and explicit memory for a motor sequence. *Psychology and Aging, 7* (3), 425-434.

Hsu, S.H., Huang, C.C., Tsuang, Y.H., & Sun, J.S. (1997). Age differences in remote pointing performance. *Perceptual Motor Skills, 85,* 2515-2527.

Izquierdo, M., Aguado, X., Gonzalez, R., Lopez, J.L., & Hakkinen, K. (1999). Maximal and explosive force production capacity and balance performance in men of different ages. *European Journal of Applied Physiology, 79,* 260-267.

Judge, J.O., King, M.B., Whipple, R., Clive, J., & Wolfson, L.I. (1995). Dynamic balance in older persons: effects of reduced visual and proprioceptive input. *Journal of Gerontology, 50A* (5), M263-M270.

Ketcham, C.J., Dounskaia, N.V., and Stelmach, G.E. (2004). Age-related differences in the control of multijoint movements. *Motor Control, 8,* 422-436.

Ketcham, C.J., Seidler, R.D., Van Gemmert, A.W.A., & Stelmach, G.E. (2002). Age related kinematic differences as influenced by task difficulty, target-size, and movement amplitude. *Journal of Gerontology: Psychological Sciences and Social Sciences, 57B,* P54-P64.

Ketcham, C.J., & Stelmach, G.E. (2002). Motor control of older adults. In D.J. Ekerdt, R.A. Applebaum, K.C. Holden, S.G. Post, K. Rockwood, R. Schulz, R.L. Sprott, & P. Uhlenberg (Eds.), *Encyclopedia of Aging.* New York: Macmillan Reference USA.

Kinoshita, H., & Francis, P.R. (1996). A comparison of prehension force control in young and elderly individuals. *European Journal of Applied Physiology, 74,* 450-460.

Kramer, A.F., Hahn, S., & Gopher, D. (1999). Task coordination and aging: explorations of executive control processes in the task switching paradigm. *Acta Psychologica, 101* (2-3), 339-378.

Larish, D.D., & Stelmach, G.E. (1982). Preprogramming, programming, and reprogramming of aimed hand movements as function of age. *Journal of Motor Behavior, 14* (4), 322-340.

Larsson, L., & Karlsson, J. (1978). Isometric and dynamic endurance as a function of age and skeletal muscle characteristics. *Acta Physiologica Scandinavica, 104,* 129-136.

Lazarus, J.C., & Haynes, J.M. (1997). Isometric pinch force control and learning in older adults. *Experimental Aging Research, 23,* 179-200.

Marteniuk, R.G., MacKenzie, C.L., Jeannerod, M., Athenes, S., & Dugas, C. (1987). Constraints of human arm movement trajectories. *Canadian Journal of Psychology, 41* (3), 365-378.

Meyer, D.E., Abrams, R.A., Kornblum, S., Wright, C.E., & Smith, J.E.K. (1988). Optimality in human motor performance: ideal control of rapid aimed movements. *Psychological Review, 95,* 340-370.

Milner, T.E., Cloutier, C., Leger, A.B., & Franklin, D.W. (1995). Inability to activate muscles maximally during cocontraction and the effect on joint stiffness. *Experimental Brain Research, 107,* 293-305.

Milner-Brown, H.S., Stein, R.B., & Yemm, R. (1973). The contractile properties of human motor units during voluntary isometric contractions. *Journal of Physiology, 228* (2), 285-306.

Morgan, M., Phillips, J.G., Bradshaw, J.L., Mattingly, J.B., Iasek, R., & Bradshaw, J.A. (1994). Age-related motor slowness: simply strategic? *Journal of Gerontology, 49A* (2), M133-M139.

Murrell, F.H. (1970). The effect of extensive practice on age differences in reaction time. *Journal of Gerontology, 25,* 268-274.

Ng, A.V., & Kent-Braun, J.A. (1999). Slowed muscle contractile properties are not associated with a decreased EMG/force relationship in older humans. *Journal of Gerontology, 54A* (10), B452-B458.

Peterka, R.J., & Black, F.O. (1990). Age-related changes in human posture control sensory organization tests. *Journal of Vestibular Research, 1,* 73-85.

Pohl, P.S., Winstein, C.J., & Fisher, B.E. (1996). The locus of age-related movement slowing: sensory processing in continuous goal-directed aiming. *Journal of Gerontology, 51B* (2), P94-P102.

Pratt, J., Chasteen, A.L., & Abrams, R.A. (1994). Rapid aimed limb movements: Age differences and practice effects in component submovements. *Psychology and Aging, 9* (2), 325-334.

Proteau, L., Charest, I., & Chaput, S. (1994). Differential roles with aging of visual and proprioceptive afferent information for fine motor control. *Journal of Gerontology, 49B* (3), P100-P107.

Romero, D.H., Van Gemmert, A.W.A., Adler, C.H., Bekkering, H., & Stelmach, G.E. (2003). Time delays prior to movement alter the drawing kinematics of elderly adults. *Human Movement Science, 22* (2), 207-220.

Roos, M.R., Rice, C.L., Connelly, D.M., & Vandervoort, A.A. (1999). Quadriceps muscle strength contractile properties, and motor unit firing rates in young and old men. *Muscle and Nerve, 22* (8), 1094-1103.

Rose, D.J., & Clark, S. (2000). Can the control of bodily orientation be significantly improved in a group of older adults with a history of falls? *Journal of the American Geriatrics Society, 48,* 275-282.

Salthouse, T.A. (1984). Effects of age and skill in typing. *Journal of Experimental Psychology: General, 113,* 345-371.

Salthouse, T.A. (1985). A theory of cognitive aging. In G.E. Stelmach & P.A. Vroon (Eds.), *Advances in Psychology, 28.* Amsterdam: Elsevier Science BV (North-Holland).

Salthouse, T.A., & Somberg, B.L. (1982). Isolating the age deficit in speeded performance. *Journal of Gerontology, 37,* 59-63.

Seidler, R.D., Alberts, J.L., & Stelmach, G.E. (2002). Changes in multi-joint performance with age. *Motor Control, 6,* 19-31.

Seidler, R.D., & Stelmach, G.E. (1995a). Reduction in control with aging: temporal and spatial declines. In S.N. Blair (Ed.), *Physical Activity, Fitness, and Health.* Champaign, IL: Human Kinetics.

Seidler, R.D., & Stelmach, G.E. (1995b). Reduction in sensorimotor control with age. *Quest, 47,* 386-394.

Seidler-Dobrin, R.D., He, J., & Stelmach, G.E. (1998). Coactivation to reduce variability in the elderly. *Motor Control, 2,* 314-330.

Seidler-Dobrin, R.D., & Stelmach, G.E. (1998). Persistence in visual feedback control by the elderly. *Experimental Brain Research, 119,* 467-474.

Singh, M.A.F., Ding, W., Manfredi, T.J., Solares, G.S., O'Neill, E.F., Clements, K.M., Ryan, N.D., Kehayias, J.J., Fielding, R.A., & Evans, W.J. (1999). Insulin-like growth factor I in skeletal muscle after weight-lifting exercise in frail elders. *American Journal of Physiology, 277 (Endocrinology and Metabolism, 40)*, E135-E143.

Slavin, M.J., Phillips, J.G., & Bradshaw, J.L. (1996). Visual cues in the handwriting of older adults: a kinematic analysis. *Psychology and Aging, 11* (3), 521-526.

Spirduso, W. (1995). *Physical Dimensions of Aging.* Champaign, IL: Human Kinetics.

Stelmach, G.E., & Goggin, N.L. (1988). Psychomotor decline with age. *Physical Activity and Aging, 22,* 6-17.

Stelmach, G.E., & Sirica, A. (1986). Aging and proprioception. *Age, 9,* 99-103.

Strayer, D.L., & Kramer, A.F. (1994). Aging and skill acquisition: learning-performance distinctions. *Psychology and Aging, 9* (4), 589-605.

Swinnen, S.P., Verschueren, S.M.P., Bogaerts, H., Dounskaia, N., Lee, T.D., Stelmach, G.E., & Serrien, D.J. (1998). Age-related deficits in motor learning and differences in feedback processing during the production of a bimanual coordination pattern. *Cognitive Neuropsychology, 15* (5), 439-466.

Teulings, H.L., & Stelmach, G.E. (1993). Signal-to-noise ratio of handwriting size, force, and time: cues to early markers of Parkinson's disease. In G.E. Stelmach & V. Homberg (Eds.), *Sensorimotor Impairment in the Elderly* (pp. 311-327). Amsterdam: Elsevier Science BV (North-Holland).

Walker, N., Philbin, D.A., & Fisk, A.D. (1997). Age-related differences in movement control: adjusting submovement structure to optimize performance. *Journal of Gerontology, 52B* (1), P40-P52.

Warabi, T., Noda, H., & Kato, T. (1986). Effect of aging on sensorimotor functions of eye and hand movements. *Experimental Neurobiology, 93,* 686-697.

Welford, A.T. (1977). Motor performance. In J.E. Birren & K.W. Schaie (Eds.), *Handbook for the Psychology of Aging.* New York: Van Nostrand Reinhold.

Welford, A.T. (1984). Between bodily changes and performance: some possible reasons for slowing with age. *Experimental Aging Research, 10* (2), 73-88.

Whipple, R., Wolfson, L., Derby, C., Singh, D., & Tobin, J. (1993). Altered sensory function and balance in older persons. *Journal of Gerontology, 48* (SI), 71-76.

Wishart, L.R., Lee, T.D., Murdoch, J.E., & Hodges, N.J. (2000). Effects of aging on automatic and effortful processes in bimanual coordination. *Journal of Gerontology, 53B* (2), P85-P94.

Woollacott, M.H. (1993). Age-related changes in posture and movement. *Journal of Gerontology, 48* (SI), 56-60.

Yan, J.H., Thomas, J.R., & Stelmach, G.E. (1998). Aging and rapid aiming arm movement control. *Experimental Aging Research, 24,* 155-168.

Yue, G.H., Ranganathan, V.K., Siemionow, V., Liu, J.Z., & Sahgal, V. (1999). Older adults exhibit a reduced ability to fully activate their biceps brachii muscle. *Journal of Gerontology, 54A* (4), M249-M253.

Chapter 5

Cohen, U., & Moore, K.D. (1999). Integrating cultural heritage into assisted living environments. In B. Schwarz & R. Brent (Eds.), *Aging, Autonomy and Architecture: Advances in Assisted Living.* Baltimore: Johns Hopkins University Press.

Cohen, U., & Weisman, G. (1991). *Holding on to Home: Designing Environments for People with Dementia.* Baltimore: Johns Hopkins University Press.

Day, K., & Cohen, U. (2000). The role of culture in designing environments for people with dementia: a study of Russian Jewish immigrants. *Environment and Behavior, 32* (3), 361-399.

Chapter 6

American Cancer Society. (2005). *Cancer Facts & Figures 2003.* Accessed April 8, 2005, from www.cancer.org/downloads/STT/CAFF2005f4PWSecured.pdf.

Bandura, A. (1986). *Social Foundations of Thought and Action: A Social Cognitive Theory.* Englewood Cliffs, NJ: Prentice Hall.

Blumenthal, J.A., Emery, C.F., Madden, D.J., Schniebolk, S., Walsh-Riddle, M., George, L.K., McKee, D.C., Higginbotham, M.B., Cobb, F.R., & Coleman, R.E. (1991). Long-term effects of exercise on psychological functioning in older men and women. *Journal of Gerontology, 46* (6), P352-361.

Blumenthal, J.A., & Madden, D.J. (1988). Effects of aerobic exercise training, age, and physical fitness on memory-search performance. *Psychology of Aging, 3* (3), 280-285.

Bryk, A., & Raudenbush, S.W. (1992). *Hierarchical Linear Models for Social and Behavioral Research: Applications and Data Analysis Methods.* Newbury Park, CA: Sage.

Cella, D.F., Eton, D.T., Lai, J.-S., Peterman, A.H., & Merkel, D.E. (2002). Combining anchor and distribution-based methods to derive minimal clinically important differences on the Functional Assessment of Cancer Therapy (FACT) Anemia and Fatigue Scales. *Journal of Pain and Symptom Management, 24* (6), 547-561.

Courneya, K.S., & Friedenreich, C.M. (1999). Physical exercise and quality of life following cancer diagnosis: a literature review. *Annals of Behavioral Medicine, 21* (2), 171-179.

Diener, E. (1984). Subjective well-being. *Psychological Bulletin, 95* (3), 542-575.

Diener, E., Emmons, R.A., Larsen, R.J., & Griffin, S. (1985). The Satisfaction With Life Scale. *Journal of Personality Assessment, 49* (1), 71-75.

Duncan, T.E., Duncan, S.C., Strycker, L.A., Fuzhong, L., & Alpert, A. (1999). *An Introduction to Latent Variable Growth Curve Modeling: Concepts, Issues, and Applications.* Mahwah, NJ: Erlbaum.

Garratt, A., Schmidt, L., Mackintosh, A., & Fitzpatrick, R. (2002). Quality of life measurement: bibliographic study of patient assessed health outcome measures. *British Medical Journal, 324* (7351), 1417-1421.

Jerome, G.J., Marquez, D.X., McAuley, E., Canaklisova, S., Snook, E., & Vickers, M. (2002). Self-efficacy effects on feeling states in women. *International Journal of Behavioral Medicine, 9,* 139-154.

Keysor, J.J. (2003). Does late-life physical activity or exercise prevent or minimize disablement?: a critical review of the scientific evidence. *American Journal of Preventive Medicine, 25* (3, Suppl. 2), 129-136.

Marquez, D.X., Jerome, G.J., McAuley, E., Snook, E., & Canaklisova, S. (2002). Self-efficacy manipulation and state anxiety responses to exercise in low active women. *Psychology and Health, 17* (6), 783-791.

Masse, L.C., Dassa, C., Gauvin, L., Giles-Corti, B., & Motl, R. (2002). Emerging measurement and statistical methods in physical activity research. *American Journal of Preventive Medicine, 23* (2S), 44-55.

McAuley, E., Blissmer, B., Marquez, D.X., Jerome, G.J., Kramer, A.F., & Katula, J. (2000). Social relations, physical activity, and well-being in older adults. *Preventive Medicine, 31,* 608-617.

McAuley, E., & Katula, J. (1998). Physical activity interventions in the elderly: influence on physical health and psychological function. In R. Schulz, M.P. Lawton, & G. Maddox (Eds.), *Annual Review of Gerontology and Geriatrics* (pp. 115-154). New York: Springer.

McAuley, E., & Rudolph, D. (1995). Physical activity, aging, and psychological well-being. *Journal of Aging and Physical Activity, 3,* 67-96.

McAuley, E., Talbot, H.-M., & Martinez, S. (1999). Manipulating self-efficacy in the exercise environment in women: influences on affective responses. *Health Psychology, 18* (3), 288-294.

Miller, M.E., Rejeski, W.J., Reboussin, B.A., Ten Have, T.R., & Ettinger, W.H. (2000). Physical activity, functional limitations, and disability in older adults. *Journal of the American Geriatrics Society, 48* (10), 1264-1272.

Muthén, B.O. (2001). Second-generation structural equation modeling with combination of categorical and continuous latent variables: new opportunities for latent class/latent growth modeling. In L.M. Collins & A.G. Sayer (Eds.), *New Methods for the Analysis for Change* (pp. 291-322). Washington, DC: American Psychological Association.

Pavot, W., & Diener, E. (1993). The affective and cognitive context of self-reported measures of subjective well-being. *Social Indicators Research, 28* (1), 1-20.

Rejeski, W.J., Brawley, L.R., & Shumaker, S.A. (1996). Physical activity and health-related quality of life. *Exercise and Sport Sciences Reviews, 24,* 71-108.

Rejeski, W.J., & Mihalko, S.L. (2001). Physical activity and quality of life in older adults. *Journal of Gerontology: Biological Sciences and Medical Sciences, 56A* (Special Issue No. 2), 23-35.

Schechtman, K.B., & Ory, M.G. (2001). The effects of exercise on the quality of life of frail older adults: a preplanned meta-analysis of the FICSIT trials. *Annals of Behavioral Medicine, 23* (3), 186-197.

Singh, M.A. (2002). Exercise to prevent and treat functional disability. *Clinical Geriatric Medicine, 18* (3), 431-462, vi-vii.

Spirduso, W.W., & Cronin, D.L. (2001). Exercise dose-response effects on quality of life and independent living in older adults. *Medicine and Science in Sports and Exercise, 33* (6 Suppl.), S598-608.

Stewart, A.L., & King, A.C. (1991). Evaluating the efficacy of physical activity for influencing quality-of-life outcomes in older adults. *Annals of Behavioral Medicine, 13* (3), 108-116.

Stewart, A.L., Mills, K.M., Sepsis, P.G., King, A.C., McLellan, B.Y., Roitz, K., & Ritter, P.L. (1997). Evaluation of CHAMPS, a physical activity promotion program for older adults. *Annals of Behavioral Medicine, 19* (4), 353-361.

Thomas, D.R. (2001). Critical link between health-related quality of life and age-related changes in physical activity and nutrition. *Journal of Gerontology: Medical Sciences, 56A* (10), M599-M602.

Trine, M.R. (1999). Physical activity and quality of life. In J.M. Rippe (Ed.), *Lifestyle Medicine* (pp. 989-997). Malden, MA: Blackwell Science.

U.S. Census Bureau. (2000). *Population Profile of the United States: 2000 (Internet Release)*. Accessed November 25, 2003, from www.census.gov/population/pop-profile/2000/profile2000.pdf.

Ware, J.E., & Sherbourne, C.D. (1992). The MOS 36-item short-form health survey (SF-36). I. Conceptual framework and item selection. *Medical Care, 30* (6), 473-483.

Yellen, S.B., Cella, D.F., Webster, K., Blendowski, C., & Kaplan, E. (1997). Measuring fatigue and other anemia-related symptoms with the Functional Assessment of Cancer Therapy (FACT) measurement system. *Journal of Pain and Symptom Management, 13* (2), 63-74.

Chapter 7

Coyle, C.P., Santiago, M.C., Shank, J.W., Ma, G.X., & Boyd, R. (2000). Secondary conditions and women with physical disabilities: a descriptive study. *Archives of Physical Medicine and Rehabilitation, 81,* 1380-1387.

Durante, R., & Ainsworth, B.E. (1996). The recall of physical activity: using a cognitive model of the question-answering process. *Medicine and Science in Sports and Exercise, 28* (10), 1282-1291.

Durstine, J.L., Painter, P.P., Franklin, B.A., Morgan, D., Pitetti, K.H., & Roberts, S.O. (2000). Physical activity for the chronically ill and disabled. *Sports Medicine, 30,* 207-219.

Freedson, P.S., Melanson, E., & Sirard, J. (1998). Calibration of the Computer Science and Applications, Inc. accelerometer. *Medicine and Science in Sports and Exercise, 30,* 777-781.

Fujiura, G.T. (2001). Emerging issues in disability demographics. *Population Today, 29,* 9-10.

Gordon, J.S., Heil, D.P., & Bauer, M.J. (1999). Energy expenditure prediction accuracy of the CSA accelerometer for overground walking. *Medicine and Science in Sports and Exercise, 31* (Suppl.), S143.

Heath, G.W., & Fentem, P.H. (1997). Physical activity among persons with disabilities public health perspective. *Exercise and Sport Sciences Reviews, 25,* 195-234.

Kinne, S., Patrick, D.L., & Maher, E.J. (1999). Correlates of exercise maintenance among people with mobility impairments. *Disability and Rehabilitation, 21,* 15-22.

Osness, W.H., Adrian, M., Clark, B., Hoeger, W., Raab, D., & Wiswell, R. (1996). *Functional Fitness Assessment for Adults over 60 Years*. Dubuque, IA: Kendall/Hunt.

Patrick, D.L., Richardson, M., Starks, H.E., & Rose, M.A. (1994). A framework for promoting the health of people with disabilities. In D. Lollar (Ed.), *Preventing Secondary Conditions Associated with Spina Bifida or Cerebral Palsy: Proceedings and Recommendations of a Symposium.* (pp. 3-16). Washington, DC: Spina Bifida Association of America.

Ravesloot, C., Seekins, T., & Young, Q. (1998). Health promotion for people with chronic illness and physical disabilities: the connection between health psychology and disability prevention. *Clinical Psychology and Psychotherapy, 5,* 76-85.

Rejeski, J.W., & Focht, B. (2002). Aging and physical disability: on integrating group and individual counseling with the promotion of physical activity. *Exercise and Sports Science Reviews, 30,* 166-170.

Rice, M.W., & Trupin, L. (1996). Medical expenditures for people with disabilities. *Disability Statistics Abstracts, 12,* 1-4.

Rikli, R.E., & Jones, J. (1999). Development and validation of a functional fitness test for community residing older adults. *Journal of Aging and Physical Activity, 7,* 129-161.

Rimmer, J.H. (1999). Health promotion for people with disabilities: the emerging paradigm shift from disability prevention to prevention of secondary conditions. *Physical Therapy, 79* (5), 495-502.

Rimmer, J.H. (2001). Resistance training for persons with physical disabilities. In J.E. Graves & B.A. Franklin (Eds.), *Resistance Training for Health and Rehabilitation* (pp. 321-346). Champaign, IL: Human Kinetics.

Rimmer, J.H., & Braddock, D. (1997). Physical activity, disability, and cardiovascular health. In A.S. Leon (Ed.), *Physical Activity and Cardiovascular Health. A National Consensus* (pp. 236-244). Champaign, IL: Human Kinetics.

Rimmer, J.H., Braddock, D., & Pitetti, K.H. (1996). Research on physical activity and disability: an emerging national priority. *Medicine and Science in Sports and Exercise, 28,* 1366-1372.

Rimmer, J.H., Riley, B.B., & Rubin, S.S. (2001). A new measure for assessing the physical activity behaviors of persons with disabilities: The Physical Activity and Disability Survey. American Journal of Health Promotion, 16, 34-45.

Rimmer, J. H., Riley, B., Wang, E., & Rauworth, A. (2004). Development and validation of AIMFREE: Accessibility Instruments Measuring Fitness and Recreation Environments. *Disability & Rehabilitation, 26,* 1087-1095.

Rimmer, J. H., Riley, B., Wang, E., Rauworth, A., & Jurkowski, J. (2004). Physical activity participation among persons with disabilities: Barriers and facilitators. *American Journal of Preventive Medicine, 26,* 419-425.

Tortolero, S.R., Masse, L.C., Fulton, J.E., Torres, I., & Kohl, H.W. (1999). Assessing physical activity among minority women: focus group results. *Women's Health Issues, 9,* 135-142.

Trost, S.G. (2001). Objective measurement of physical activity in youth: current issues, future directions. *Exercise and Sport Sciences Reviews, 29,* 32-36.

U. S. Department of Health and Human Services, Public Health Service, Centers for Disease Control and Prevention, National Center for Chronic Disease Prevention and Health Promotion, Division of Nutrition and Physical Activity. (1999). *Promoting physical activity. A guide for community action.* Champaign, IL: Human Kinetics.

U.S. Department of Health and Human Services. (2000). *Healthy People 2010* (chapter 6, pp. 6-1–6-28). Washington, DC: U.S. Department of Health and Human Services.

World Health Organization. (2001). *International Classification of Functioning, Disability and Health.* Geneva: World Health Organization.

Zola, I.K. (1993). Disability statistics. What we count on and what it tells us: a personal and political analysis. *Journal of Disability Policy Studies, 4,* 9-39.

Chapter 8

Clark, D.O. (1997). Physical activity efficacy and effectiveness among older adults and minorities. *Diabetes Care, 20,* 1176-1182.

Dishman, R.K. (1994). *Advances in Exercise Adherence.* Champaign, IL: Human Kinetics.

Kenney, W.L. (1993). The older athlete: exercise in hot environments. *Sport Science Exchange #44, 6* (3).

Marcus, B., & Forsyth, L.H. (2003). *Motivating People to Be Physically Active.* Champaign, IL: Human Kinetics.

McAuley, E., Blissmer, B., Katula, J., Duncan, T.E., & Mihalko, S.L. (2000). Physical activity, self-esteem, and self-efficacy relationships in older adults: a randomized controlled trial. *Annals of Behavioral Medicine, 22,* 131-139.

McAuley, E., Blissmer, B., Marquez, D.X., Jerome, G.J., Kramer, K.F., & Katula, J. (2000), Social relations, physical activity, and well-being in older adults. *Preventive Medicine, 31,* 608-617.

Moudon, A.V., & Lee, C. (2003). Walking and biking: an evaluation of environmental audit instruments. *American Journal of Health Promotion, 19,* 21-37.

Saelens, B.E., Sallis, J.F., Black, J.B., & Chen, D. (2003). Neighborhood-based differences in physical activity: an environment scale evaluation. *American Journal of Public Health, 93,* 1552-1558.

Thompson P.D., Buchner, D., Pina, I.L., Balady, G.J., Williams, M.A., Marcus, B.H., Berra, K., Blair, S.N., Costa, F., Franklin. B., Fletcher, G.F., Gordon, N.F., Pate, R.R., Rodriguez, B.L., Yancey, A.K., & Wenger, N.K.; American Heart Association Council on Clinical Cardiology Subcommittee on Exercise, Rehabilitation, and Prevention; American Heart Association Council on Nutrition, Physical Activity, and Metabolism Subcommittee on Physical Activity. (2003). Exercise and physical activity in the prevention and treatment of atherosclerotic cardiovascular disease: a statement from the Council on Clinical Cardiology (Subcommittee on Exercise, Rehabilitation, and Prevention) and the Council on Nutrition, Physical Activity, and Metabolism (Subcommittee on Physical Activity). *Circulation, 107,* 3109-3116.

U.S. Department of Health and Human Services. (1996). *Physical Activity and Health: A Report of the Surgeon General.* Atlanta: U.S. Department of Health and Human Services, Centers for Disease Control and Prevention, National Center for Chronic Disease Prevention and Health Promotion.

Chapter 9

Agency for Healthcare Research and Quality and Centers for Disease Control (AHRQ/CDC). (2002). *Physical Activity and Older Americans: Benefits and Strategies.* Accessed June 20, 2002, from www.ahrq.gov/ppip/activity/htm.

American Association of Retired Persons. (2003). Be active for life. Accessed October 10, 2003, from www.aarp.org/activeforlife/.

American College of Sports Medicine. (1998). Position Stand. Exercise and physical activity for older adults. *Medicine and Science in Sports and Exercise, 30,* 992-1008.

Booth, F.W., & Chakravarthy, M.V. (2002). Cost and consequences of sedentary living: new battleground for an old enemy. *President's Council on Physical Fitness and Sports Research Digest, 3,* 16.

Bush, G.W. (2002). *Healthier US: The President's Health and Fitness Initiative.* Executive Summary. www.whitehouse.gov/infocus/fitness/execsummary.html.

Center on an Aging Society. (2003). *Obesity Among Older Americans.* Washington, DC: Georgetown University, Institute for Health Care Research and Policy.

Centers for Disease Control and Prevention (2004). Behavioral Risk Factor Surveillance System. Prevalence and Trend Data. Accessed April 14, 2005, from http://apps.nccd.cdc.gov/brfss.

Evashwick, C., & Ory, M.G. (2003). Organizational characteristics of successful innovative programs sustained over time. *Journal of Family and Community Health, 26* (3), 177-193.

Glasgow, R.E., Vogt, T.M., & Boles, S.M. (1999). Evaluating the public health impact of health promotion interventions: the RE-AIM framework. *American Journal of Public Health, 89,* 1322-1327.

Jette, A.M., Lachman, M., Giorgetti, M.M., Assmann, S.F., Harris, B.A., Levenson, C., Wernick, M., and Krebs, D. (1999). Exercise—it's never too late: the strong-for-life program. *American Journal of Public Health, 89,* 66-72.

King, A.C., Bauman, A., & Calfas, K. (Guest Eds.). (2002). Innovative approaches to understanding and influencing physical activity. *American Journal of Preventive Medicine, 23* (Suppl. 2), 1-108.

King, A.C., Castro, C., Wilcox, S., Eyler, A.A., Sallis, J.F., & Brownson, R.C. (2000). Personal and environmental factors associated with physical inactivity among different racial/ethnic groups of U.S. middle and older-aged women. *Health Psychology, 19,* 354-364.

King, A.C., Rejeski, W.J., & Buchner, D.M. (1998). Physical activity interventions targeting older adults: a critical review. *American Journal of Preventive Medicine, 15,* 316-333.

McLeroy, K.R., Bibeau, D., Steckler, A., & Glanz, K. (1988). An ecological perspective on health promotion programs. *Health Education Quarterly, 15,* 351-377.

National Council on Aging. (2001). A National Survey of Health and Supportive Services in the Aging Network. Accessed April 14, 2005, from http://www.ncoa.org/Downloads/cbo_report.pdf.pdf.

National Institute on Aging. (2001). Exercise: A Guide from the National Institute on Aging (NIH Publication No. 01-4258). Gaithersburg, MD: National Institutes of Health.

Oldenburg, B., & Parcel, G.S. (2002). Diffusion of innovations. In K. Glanz, B.K. Rimer, & F.M. Lewis (Eds.), *Health Behavior and Health Education* (pp. 312-334). San Francisco: Jossey-Bass.

Ory, M.G., Hoffman, M., Hawkins, M., Sanner, B., & Mockenhaupt, R. (2003). Challenging aging stereotypes: strategies for creating a more active society. *American Journal of Preventive Medicine, 25* (3S2), 164-171.

Ory, M.G., Jordan, P.J., & Bazzarre, T. (2002). The Behavior Change Consortium: setting the stage for a new century of health behavior-change research. *Health Education Research, 17* (5), 500-511.

Pate, R.R., Pratt, M., Blair, S.N., Haskell, W.L., Macera, C.A., Bouchard, C., Buchner, D., Ettinger, W., Heath, G.W., and King, A.C. (1995). Physical activity and public health. A recommendation from the Centers for Disease Control and Prevention and the American College of Sports Medicine. *Journal of the American Medical Association, 273,* 402-407.

Robert Wood Johnson Foundation. (2001). National blueprint for increasing physical activity among adults 50 and older: creating a strategic framework and enhancing organizational capacity for change. *Journal of Aging and Physical Activity, 9* (Suppl.), S5-S28.

Rogers, E.M. (1995). *Diffusion of Innovations* (4th ed.). New York: Free Press.

Sallis, J.F., & Owen, N. (2002). Ecological models. In K. Glanz, B.K. Rimer, & F.M. Lewis (Eds.), *Health Behavior and Health Education: Theory, Research, and Practice* (3rd ed., pp. 462-484). San Francisco: Jossey-Bass.

Smedley, B.D., Syme, S.L., eds. (2000). *Promoting Health: Intervention Strategies from Social and Behavioral Research.* Institute of Medicine. Washington, DC: National Academy Press.

Stokols, D. (1996). Translating social ecological theory into guidelines for community health promotion. *American Journal of Health Promotion, 10,* 282-298.

U.S. Department of Health and Human Services. (2000). Healthy People 2010: Leading Health Indicators. Accessed April 14, 2005, from http://www.healthypeople.gov/LHI/.

U.S. Department of Health and Human Services. (1996). *Physical activity and health: a report of the Surgeon General.* Atlanta, Georgia: U.S. Department of Health and Human Services, Public Health Service, CDC, National Center for Chronic Disease Prevention and Health Promotion.

U.S. Department of Health and Human Services. (1999). *Promoting Physical Activity: A Guide for Community Action.* Atlanta, Georgia: U.S. Department of Health and Human Services, Centers for Disease Control and Prevention, National Center for Chronic Disease Prevention and Health Promotion, Division of Nutrition and Physical Activity.

U.S. Department of Health and Human Services. (2003). Steps to a healthier U.S. *The Power of Prevention.* Accessed April 14, 2005, from http://www.healthierus.gov/steps/summit/prevportfolio/Power_Of_Prevention.pdf.

Weiss, H., Coffman, J., & Bohan-Baker, M. (2002). Evaluation's role in supporting initiative and sustainability. *Harvard Family Research Project.* Accessed April 15, 2005, from http://www.gse.harvard.edu/hfrp/pubs/onlinepubs/sustainability/track.html.

Chapter 10

Chen, K., & Yeung, R. (2002). Exploratory studies of Qigong therapy for cancer in China. *Integrative Cancer Therapies, 1* (4), 345-370.

Iwao, M., Kajiyama, S., Mori, H., & Oogaki, K. (1999). Effects of Qigong walking on diabetic patients: a pilot study. *Journal of Alternative and Complementary Medicine, 5* (4), 353-358.

Jones, B.M. (2001). Changes in cytokine production in healthy subjects practicing Guolin Qigong: a pilot study. *BMC Complementary and Alternative Medicine, 1,* 8.

Lee, M., Huh, H., Kim, B., Ryu, H., Lee, H., Kim, J., & Chung, H. (2002). Effects of Qi-training on heart rate variability. *American Journal of Chinese Medicine, 30* (4), 463-470.

Lee, M., Kim, B., Huh, H., Ryu, H., Lee, H., & Chung, H. (2000). Effects of Qi-training on blood pressure, heart rate and respiration rate. *Clinical Physiology, 20* (3), 173-176.

Lee, M., Lee, M., & Kim, H. (2003). Qigong reduced blood pressure and catecholamine levels of patients with essential hypertension. *International Journal of Neuroscience, 113,* 1691-1701.

Li, M. (1985). Qigong: its origin and development. In *China Sports* magazine (Compiled). *The Wonders of Qigong: A Chinese Exercise for Fitness, Health, and Longevity* (pp. 11-15). Los Angeles: Wayfarer.

Lin, H. (1987). *Yang sheng qigong xue* [Study of Qigong for wellness]. Guang Zhou, China: Guang Dong Scientific.

Lin, Z., & Chen, K. (2002). Exploratory studies of external Qi in China. *Journal of International Society of Life Information Science, 20* (2), 457-461.

Lin, Z., Yu, L., Guo, Z., Shen, Z., Zhang, H., & Zhang, T. (2000). *Qigong: Chinese Medicine or Pseudoscience?* New York: Prometheus Books.

Litscher, G., Wenzel, G., Niederwieser, G., & Schwarz, G. (2001). Effects of Qigong on brain function. *Neurological Research, 23,* 501-505.

Liu, G.L., Cui, R.Q., Li, G.Z., & Huang, C.M. (1990). Changes in brainstem and cortical auditory potentials during Qigong meditation. *American Journal of Chinese Medicine, 18* (3-4), 95-103.

Liu, M., Zhang, Y., & Liu, B. (1992). *Zhong Hua Qi Gong: Shao Lin Nei Jin Yi Chi Chan* [Chinese Qi-gong: Shao Lin Nei Jin Yi Chi Chan routine]. Beijing: China International Broadcast.

Mayer, M. (1999). Qigong and hypertension: a critique of research. *Journal of Alternative and Complementary Medicine, 5* (4), 371-382.

Ng, B. (1998). Qigong-induced mental disorders: a review. *Australian and New Zealand Journal of Psychiatry, 33,* 197-206.

Sancier, K. (1996). Medical applications of Qigong. *Alternative Therapies in Health and Medicine, 2* (1), 40-47.

Sancier, K., & Holman, D. (2004). Commentary: multifaceted health benefits of medical Qigong. *Journal of Alternative and Complementary Medicine, 10* (1), 163-165.

Tsang, H.W.H., Cheung, L., & Lak, D.C.C. (2002). Qigong as a psychosocial intervention for depressed elderly with chronic physical illnesses. *International Journal of Geriatric Psychiatry, 17,* 1146-1154.

Tsang, H.W.H., Mok, C.K., Au Yeung, Y.T., & Chan, S.Y.C. (2003). The effect of Qigong on general and psychosocial health of elderly with chronic physical illnesses: a randomized clinical trial. *International Journal of Geriatric Psychiatry, 18,* 441-449.

Wu, W., Bandilla, E., Ciccone, D.S., Yang, J., Cheng, S.S., Carner, N., Wu, Y., & Shen, R. (1999). Effects of Qigong on late-stage complex regional pain syndrome. *Alternative Therapies in Health and Medicine, 5* (1), 45-54.

Yang, J.M. (1992). *The Root of Chinese Chi Kung: The Secrets of Chi Kung Training.* Jamaica Plain, MA: YMAA Publication Center.

Zhang, M., & Sun, X. (Compiled, 1985). *Chinese Qigong Therapy.* Translated by X. Yang & X. Yao. Jinan, China: Shandong Science and Technology Press.

Chapter 11

Browne, M.W. (1993). Structured latent curve models. In C.M. Cuadras & C.R. Rao (Eds.), *Multivariate Analysis: Future Directions 2* (pp. 171-197). North-Holland Series in Statistics & Probability, Vol. 7. Amsterdam: North-Holland.

Browne, M.W., & du Toit, S.H.C. (1991). Models for learning data. In L.M. Collins & J.L. Horn (Eds.), *Best Methods for the Analysis of Change* (pp. 47-68). Washington, DC: American Psychological Association.

Burt, C. (1948). The factorial analysis of temperamental traits. *British Journal of Psychology, Statistical Section, 1,* 178-203.

McArdle, J.J., & Bell, R.Q. (2002). An introduction to latent growth models for developmental data analysis. In T.D. Little, K.U. Schnabel, & J. Baumert (Eds.), *Modeling Longitudinal Data: Practical Issues, Applied Approaches, and Specific Examples* (pp. 69-107). Mahwah, NJ: Erlbaum.

McDonald, R.P. (1965). Factor analytic versus classical methods of fitting individual curves. *Perceptual and Motor Skills, 20,* 270.

McDonald, R.P. (1967). Nonlinear factor analysis. Psychometric Monograph No. 15.

McDonald, R.P. (2004). The informative analysis of individual trend curves. *Multivariate Behavioral Research*, 39, 517-563.

Meredith, W., & Tisak, J. (1990). Latent curve analysis. *Psychometrika, 55,* 107-122.

Tucker, L.R. (1958). Determination of parameters of a functional relation by factor analysis. *Psychometrika, 23,* 19-23.

Yates, A.J., & McDonald, R.P. (1966). Reminiscence as a function of perceptual search. *Australian Journal of Psychology, 18,* 137-143.

Chapter 12

Allen, C. (2003). Multi-media profiles of older people with dementia. Symposium, International Psychogeriatric Association Eleventh International Congress, Chicago. http://abstract.confex.com/ipa/11congress/techprogram/paper_3569.htm.

Baker, R. (2000). Physiotherapy, high-tech style. Inter-PACE 2000 International Foundation. www.pace2000.org/pub_citizen_19990626.html.

Bennett, R.E. (2001). How the Internet will help large-scale assessment reinvent itself. *Education Policy Analysis Archives, 9* (5). http://epaa.asu.edu/epaa/v9n5.html.

Berger, T. (2001). Brain-implantable biomemetic electronics as the next era in neural prosthetics. *Proceedings of the IEEE, 98* (7), 993-1012. http://bmsr.usc.edu/Research/pubs/vzm/993.pdf.

Campbell, C. (2003). Making stroke recovery a virtual reality. *Hospital News, 16* (6), 23-24. www.hospitalnews.com.

Czaja, S.J. (2003). Information technology for geriatric populations: a futurist view. Symposium, International Psychogeriatric Association Eleventh International Congress, Chicago. http://abstract.confex.com/ipa/11congress/techprogram/paper_2420.htm.

De Leo, D., Carollo, G., & Buono, M.D. (1995). Lower suicide rates associated with a Tele-Help/Tele-Check service for the elderly at home. *American Journal of Psychiatry, 52,* 632-634.

Fanselow, E.E., Reid, A.P., Miguel, A.L., & Nicolelis, M.A.L. (2000). Reduction of pentylenetetrazole-induced seizure activity in awake rats by seizure-triggered trigeminal nerve stimulation. *Journal of Neuroscience, 20* (21), 8160-8168.

Freitas, R.A. (2002). The future of nanofabrication and molecular scale devices in nanomedicine. *Studies in Health Technology and Informatics, 80,* 45-59.

Goho, A.M. (2003). 10 emerging technologies that will change the world: injectable tissues engineering. *Technology Review, 106* (1), 38.

Hausdorff, J.M., Mitchell, S.L., Firtion, R., Peng, R.C.K., Cudkowicz, M.E., Wei, J.Y., & Goldberger, A.L. (1997). Altered fractal dynamics of gait: reduced stride-interval correlations with aging and Huntington's disease. *Journal of Applied Physiology, 82* (1), 262-269.

Hong, C., Becker, C.R., Huber, A., Schoepf, U.J., Ohnesorge, B., Knez, A., Brüning, R., & Reiser, M.F. (2001). ECG-gated reconstructed multi-detector row CT coronary angiography: effect of varying trigger delay on image quality. *Radiology, 220,* 712-717.

Johnson, C.E., & Danhauer, J.L. (2002). A transdisciplinary holistic approach to hearing health care. *Geriatric Times, 3* (5, September/October). www.geriatrictimes.com/g021010.html.

Kennedy, P.R., & Bakay, R.A. (1998). Restoration of neural output from a paralyzed patient by a direct brain connection. *Neuroreport, 9,* 1707-1711.

Kitwood, T. (1997). *Dementia Reconsidered: the Person Comes First.* Buckingham: Open University Press.

Kurzweil, R. (1999). *The Age of Spiritual Machines: When Computers Exceed Human Intelligence.* New York: Viking.

Lemaire, E.D., & Greene, G. (2002). Continuing education in physical rehabilitation using Internet-based modules. *Journal of Telemedicine and Telecare, 8* (1), 19-24.

Norfray, J.F., & Provenzale, J.M. (2004). Alzheimer's disease: neuropathologic findings and recent advances in imaging. *American Journal of Roentgenology, 182,* 3-13.

Pew Internet and American Life Project. (2002). Getting serious online. Pew Foundation, 1100 Connecticut Ave., Washington, DC. www.pewinternet.org/.

Sands, L.P., Phinney, A., & Katz, I.R. (2000). Monitoring Alzheimer's patients for acute changes in cognitive functioning. *American Journal of Geriatric Psychiatry, 8* (February), 47-56.

Schultz, O., Sittinger, M., Haeupl, T., & Burmester, G.R. (2000). Emerging strategies of bone and joint repair. *Arthritis Research, 2,* 433-436.

Shin, D.I., Huh, S.J., Lee, T.S., & Kim, I.Y. (2003). Web-based remote monitoring of infant incubators in the ICU. *International Journal of Medical Informatics, 71* (2-3), 151-156.

Shur, M.S., Rumvantsev, S., Gaska, R., Wei, B.Q., Vaitai, R., Ajayan, P.M., & Sinius, J. (2002). Structural and transport properties of CdS films deposited on flexible substrates. *Solid State Electronics, 46* (9), 1417-1420.

Sullivan, D.C. (2000). Biomedical Imaging Symposium: Visualizing the Future of Biology and Medicine. *Radiology, 215,* 634-638.

Tilin, A. (2002). The ultimate running machine. *Wired,* October 8. www.wired.com/wired/archive/10.08/nike.html.

Trilling, J. (2001). Selections from current literature assessment of older drivers. *Family Practice, 18* (3), 339-342.

Van Sint Jan, S. (2000). The VAKHUM Project: virtual animation of the kinematics of the human. *Cultivate Interactive, 2,* October 16.

Webb, D. (2003). When independence becomes isolation. *Friends of the Elderly, 1,* 2-8.

Chapter 14

Baumgartner, T.A. (2003). *Measurement and Evaluation Council: Past, present and future.* Paper presented at the 10th Measurement and Evaluation Symposium, Champaign, IL, October.

Baumgartner, T.A., & Safrit, M.J. (2003). A genealogy of measurement specialists in physical education and exercise science. *Measurement in Physical Education and Exercise Science, 7,* 121-127.

Godbout, P., & Schutz, R.W. (1983). Generalizability of ratings of motor performances with reference to various observational designs. *Research Quarterly for Exercise and Sport, 54,* 20-27.

Hensley, L.D., Aten, R., Baumgartner, T.A., East, W.B., Lambert, L.A., & Stillwell, J.L. (1989). A survey of grading practices in public school physical education. *Journal of Research and Development in Education, 23* (1), 37-42.

Jackson, A.W., Morrow, J.R., Jensen, R.L., Jones, N.A., & Schultes, S.S. (1996). Reliability of The Prudential FITNESSGRAM™ trunk lift test in young adults. *Research Quarterly for Exercise and Sport, 67,* 115-117.

Keating, X.D. (2003). The current often implemented fitness tests in physical education programs: problems and future directions. *Quest, 55,* 141-160.

Kulinna, P.H., Cothran, D., & Regualos, R. (2003). Development of an instrument to measure student disruptive behavior. *Measurement in Physical Education and Exercise Science, 7,* 25-41.

Locke, L.F., Silverman, S.J., & Spirduso, W.W. (2004). *Reading and Understanding Research* (2nd ed.). Thousand Oaks, CA: Sage.

Looney, M.A., & Spray, J.A. (1992). Effects of violating local independence on IRT parameter estimation for the binomial trials model. *Research Quarterly for Exercise and Sport, 63,* 356-359.

Safrit, M.J., & Wood, T.M. (1986). The Health-Related Physical Fitness Test: A tri-state survey of users and non-users. *Research Quarterly for Exercise and Sport, 57,* 27-32.

Safrit, M.J., Zhu, W.M., Costa, M.G., & Zhang, L. (1992). The difficulty of sit-ups tests: An empirical investigation. *Research Quarterly for Exercise and Sport, 63,* 277-283.

Silverman, S. (2002). *Research and professional practice: an unrequited love.* Raymond A. Weiss Research Lecture presented at the annual meeting of the American Alliance for Health, Physical Education, Recreation and Dance, San Diego, April.

Silverman, S., & Keating, X.D. (2002). A descriptive analysis of research methods classes in departments of physical education/kinesiology in the United States. *Research Quarterly for Exercise and Sport, 73,* 1-9.

Smith, R.E., Schutz, R.W., Smoll, F.L., & Ptacek, J.T. (1995). Development and validation of a multidimensional measure of sport-specific psychological skills: The Athletic Coping Skills Inventory-28. *Journal of Sport and Exercise Psychology, 17,* 379-398.

Woods, M.L., Goc Karp, G., & Feltz, D.L. (2003). Positions in kinesiology and physical education at the college or university level. *Quest, 55,* 30-50.

About the Editors

Weimo Zhu, PhD, is currently an associate professor in the department of kinesiology at the University of Illinois at Urbana-Champaign and a visiting professor at Guangzhou Institute of Physical Education and Shanghai Institute of Physical Education, both in China. His major area of research is measurement and evaluation in kinesiology.

Dr. Zhu's primary research interests are in the study and application of new measurement theories (e.g., item response theory) and models to the field of kinesiology. His research works have earned him international recognition. He served as the measurement section editor of the *Research Quarterly for Exercise and Sport* from 1999 to 2005, and he is a fellow of the American Academy of Kinesiology and Physical Education, American College of Sports Medicine, and Research consortium, AAHPERD. He is a member of the Fitnessgram/Activitygram Advisory Committee. He is also a member of the editorial board for three other journals and serves on the executive committees of several national and international professional organizations. Dr. Zhu was the chair of the Measurement and Evaluation Council, AAHPERD. Currently, Dr. Zhu is also examining the application of advanced measurement and statistical techniques to several measurement issues in the area of public health. A tangible practical application of Zhu's theoretical work has been his work in the assessment of physical activity, and he is exploring a new idea (physical activity space) and technologies (voice-recognition and automatic scoring) to solve the problems raised.

Wojtek Chodzko-Zajko, PhD, serves as both department head and professor of kinesiology at the University of Illinois at Urbana-Champaign. Dr. Chodzko-Zajko's primary research is in the area of aging and physical activity. For the past 15 years, he has focused on the effects of exercise and physical activity in the health and quality of life in older adults.

Dr. Chodzko-Zajko chairs the Active Aging Partnership, a national coalition in healthy aging linking the American College of Sports Medicine, the National Institute of Aging, the Centers for Disease Control and Prevention, the American Geriatrics Society, the National Council on the Aging, the American Association of Retired Persons, and the Robert Wood Johnson Foundation. He was editor of the *Journal of Aging and Physical Activity* and president of the International Society for Aging and Physical Activity.

Since 2002, Dr. Chodzko-Zajko has served as principal investigator of the National Blueprint Project, a coalition of more than 50 national organizations with a joint commitment to promoting independent, active aging in the older adult population.